The *ortho* Side of PCBs

Occurrence and Disposition

The *ortho* Side of PCBs

Occurrence and Disposition

by

Larry G. Hansen

University of Illinois at Urbana-Champaign

KLUWER ACADEMIC PUBLISHERS
Boston / Dordrecht / London

Distributors for North, Central and South America:
Kluwer Academic Publishers
101 Philip Drive
Assinippi Park
Norwell, Massachusetts 02061 USA
Telephone (781) 871-6600
Fax (781) 871-6528
E-Mail <kluwer@wkap.com>

Distributors for all other countries:
Kluwer Academic Publishers Group
Distribution Centre
Post Office Box 322
3300 AH Dordrecht, THE NETHERLANDS
Telephone 31 78 6392 392
Fax 31 78 6546 474
E-Mail <services@wkap.nl>

Electronic Services <http://www.wkap.nl>

Library of Congress Cataloging-in-Publication Data

Hansen, L. G.
The ortho side of PCBs : occurrence and disposition / by Larry G. Hansen.
p. cm.
Includes bibliographical references and index.
ISBN 0-7923-8541-1 (alk. paper)
1. Polychlorinated biphenyls--Toxicology. I. Title.
RA1242.P7H36 1999
615.9'512--dc21 99-27960
CIP

Printed on acid-free paper.

Printed in the United States of America

DEDICATION

To my wife, Terry, for her patience, counsel and encouragement.

To my daughter, Lori, for her strength and determination.

To family and friends who helped us all get on with our lives after the devastating assault on Lori in February, 1998.

TABLE OF CONTENTS

PREFACE ix

ACKNOWLEDGMENTS xi

1. PCBs: PRODUCTION AND PROPERTIES 1

2. THE INDIVIDUAL CHLOROBIPHENYLS 7

   2.1 GC Retention Times 7
   2.2 Other Properties 10
   2.3 Congener Abundance in Former Commercial Mixtures 14
   2.4 Congener-Specific Biological Properties 16

3. INTERPRETATION OF ANALYTICAL DATA 21

   3.1. Low Resolution GC and Pattern Recognition 22
   3.2. Transition from Low Resolution to High Resolution GC 28
   3.3. Selection of Analytes 32

4. ATMOSPHERIC TRANSPORT 35

   4.1. Global Compartments 35
   4.2 Atmospheric PCB Levels 36
   4.3. Dispersion of airborne PCBs 37
   4.4. Global Fractionation 41
   4.5. Character of Localized Emissions 44
   4.6. Accumulation of Airborne PCBs by Vegetation 47
   4.7. Other Environmental Pathways for Airborne PCBs 49
   4.8 Summary 50

5. THE HYDROSPHERE AND LITHOSPHERE 51

   5.1. Relative Magnitude of Reservoirs 51
   5.2. Movement and Fluxes of PCBs Among Reservoirs 59
   5.3. Bioconcentration of PCBs from Water 70
   5.4. Bioaccumulation from Multiple Pathway Exposures 72

6. METABOLISM AND METABOLITES 81

   6.1 Gone But Not Forgotten 81
   6.2 Mass Balance 82
   6.3 Oxidative Metabolism 83
   6.4 Phase 2 and Phase 3 Metabolism 93
   6.5. Dose, Time, Induction and Biotransformation Rates 97

7. DISPOSITION AND TOXICOKINETICS 101

7.1. Toxicokinetics 101
7.2 Very basic toxicokinetic models 103
7.3. Models of oral exposure 111
7.4. PBPK Model of early disposition 118
7.5 Toxicokinetics in Other Classes of Animals 122
7.6. Toxicokinetics and PCB Residue Profiles 131

8. FOOD CHAIN PROFILES 133

8.1. PCB profiles in fish 133
8.2. PCB profiles in marine mammals 134
8.3. Other profile-altering PCB exposures 137
8.4. Reporting of Human PCB Residues 145
8.5. Characterizing Human PCB Residues 151
8.6. Model of infant exposure to PCBs in human breast milk 154
8.7. Final considerations 159

REFERENCES 161

APPENDICES

1. Nomenclature of Polychlorinated Biphenyls 195
2. PCB Congener Weight % in Aroclors of 5 Types 205
3. Some Properties of Chlorobiphenyls 211
4. Estimation of PCB Congener Production Ratios 221
5. Comparison of Human Residues Reported by
Sample Authors Without Standards Available. 227
6. Airborne PCBs 231
7. Percent Congener Composition of PCBs in Various Plants 237
8. Dispersion in Air 241
9. Sediments and Water 243
10. PCB Residues in Fish from the Great Lakes 247
11. "Average" Fish Residue Profiles 251
12. Average Human Milk 255
13. Frequencies of Congener Occurrence in Breast Milk Surveys 259
14. Abbreviations 263

INDEX 265

PREFACE

Polychlorinated biphenyls (PCBs) have captured the attention of scientists and journalists. The literature is measured by the pound rather than the page, but a ~~clear~~ near consensus on actual risks to public health and the environment is still elusive. The discovery of the mechanism for Ah Receptor mediated toxicities in the mid 1970s diverted attention from the more prevalent nonplanar (*ortho*-rich) PCBs until recently. The renewed interest has created a need for an *ortho*-oriented sourcebook.

This is neither an introduction to PCBs nor a complete review of PCB chemistry. This book presents concepts in various stages of development based on knowledge accumulated over nearly 3 decades - sometimes through intense study, sometimes as a diversion/hobby while working on other topics. I have been told that questions raised, especially in Chapter 8, cannot be answered at present and the approaches suggested would require extensive effort and be very expensive. This, however, should not deter asking the questions. In addition, there are already large specific PCB congener data banks not yet tapped because the authors/owners aren't quite sure what to do with them. The attempts to assimilate more comprehensive data summaries may encourage release of even more "closet data."

In order to develop accurate perspectives from generalizations, the reader should have the opportunity to compare finite results from multiple sources. There is a large amount of tabular data derived from published tables, extrapolated from published figures, and/or from raw data provided by other investigators. Many of the tables are extensions of miscellaneous compilations scattered around my office for years. They just had to be titled and rendered legible. Hopefully, every reader will find at least a couple presentations of interest.

Much of the information is outside the current mainstream PCB literature, but still shaped the author's views. In order to judge the concepts, right or wrong, the critical reader may need to be reminded of these sources. In addition, some areas (whole chapters) were outside my immediate expertise. For these reasons, the bibliography is large and probably unique.

The novice may need to consult some of the references in Chapter 1 to gain a more complete understanding. Erickson (1997) thoroughly develops and discusses very important basic concepts. The expert will be bored with some of the information presented; at the same time, some important (my opinion) observations may be hidden in tiresome passages. There are reminders of previous connections so that passages can be revisited selectively.

ACKNOWLEDGMENTS

Being in a position to write this book arose from my heritage and the mentorship of Professors such as Ernie Hodgson, Frank Guthrie, Bob Metcalf and Bob Twardock. Professors Metcalf, Bill Huber and Ben Ewing provided the opportunity, through their Environmental Studies Program at UIUC, for my first graduate class in Comparative Environmental Toxicology in 1971. The roster attests to my further training by this class:

Dick Bevill (Professor and Head, Emeritus, UIUC)
Ron Borchard (Deceased; Professor Washington State University)
Joel Coates (Professor, Iowa State University)
Gary Koritz (Professor, UIUC)
Dennis Wilson (Professor, University of California-Davis)

Manuscripts from Ron (my first PhD advisee), Gary, Dennis and many subsequent outstanding students are cited in the references. Even though Dick was the main architect of the initial PCB proposal (and joined me several times in the middle of the night to assist in farrowing), his responsibilities changed and he never received the credit he deserved. Dennis was involved in so many aspects that his name was inadvertently omitted from Hansen and Welborn (1977).

Several individuals offered valuable editorial and conceptual suggestions. These were not always incorporated as suggested, so they are not responsible for poor grammar, poor judgement or errors. Many thanks to Silvana Andric, David Buchanan, Jack Cochran, Gary Koritz, Rick Larson, Donald Mackay, David Schaeffer, Sue Schantz and Petra Volmer. Karen Marley, Dave Schaeffer, Dave Vyverberg, Ted Whittem and Rex Hess expended considerable effort in translating my scribbles and crude ideas to meaningful figures.

Lori Hansen did much of the library work. Henry Anderson, Jack Cochran, Mark Hahn, Yoshihisa Kato, Debdas Mukerjee, Dave Spink and Luigi Turrio-Baldassarri provided valuable supplements. Special thanks to investigators who provided raw data, suggestions and interpretations. These include Alison Cullen, George Frame, Hannu Kiviranta, W. Harvey Newsome and Terttu Vartiainen.

Very special thanks to Shirley Pelmore for the final manuscript. Also for picking up the flow early and not being afraid to ask "Is this what you really want to say?" Even more special thanks for maintaining a pleasant, patient, understanding, and encouraging demeanor during the intervening drafts, false-starts and post-"last-time" revisions. I really don't know how she managed! I would have killed me.

1 PCBs: PRODUCTION AND PROPERTIES

Polychlorinated biphenyls (generically referred to as PCBs) were industrial compounds which enjoyed wide usage between 1930 and the mid-1970's (ATSDR, 1997; DeVoogt and Brinkman, 1989; Hutzinger et al., 1974). Desirable properties were thermal and chemical stability, low flammability, and low electrical conductance. When widespread environmental contamination by PCBs was confirmed in the late 1960s, there were concerns about human health effects and environmental impact. There is still no consensus on the significance of human health effects, but manufacture was discontinued in the US, Japan and western Europe by the late 1970's; manufacture continued in some countries until 1990. Some products containing large amounts of PCBs, especially electrical capacitors and transformers, are still in service. Other reservoirs include landfills, water and sediments. The atmosphere is an important environmental reservoir as well as a major route of exposure and transfer among other matrices.

Although many countries manufactured PCB mixtures, the most information is available for Aroclors (USA), Kanechlors (Japan) and Clophens (Germany). Some of the specific products are listed in **Table 1.1**. The less chlorinated Aroclors 1221, 1232 (a blend of 1242 and 1221) and 1016 (a distillate of Aroclor 1242) were increasingly used as replacements for the more persistent Aroclors after 1971.

Table 1.1. Characteristics of representative former commercial PCB mixtures.

Aroclor	Clophen	Kanechlor	Chlorine (%)	Chlorines (Avg No.)	Molecular Weight
1221			21	1.15	194
1232		200	32	2	223
1016			40	3	257
1242	A30	300	42	3	257.5
1248	A40	400	48	4	292
1254	A50	500	54	5	326
1260	A60	600	60	6.3	366
1262			62	6.8	388
1268			68	8.7	453.5

Actual sales figures appear to have been somewhat elusive and production estimates vary among sources. Erickson (1997) presents a reasonable assessment of the various US and global production estimates. As an example, a portion of the US sales figures (provided in graphic form as metric tons) by Hutzinger et al. (1974) are compared in **Table 1.2** to those provided for the same period by DeVoogt and Brinkman (1989) in **Table 1.3**.

Table 1.2. Domestic sales of Monsanto's PCBs in the US as presented by Hutzinger et al. (1974) (crude estimate from figure converted from metric tons to 10^6 pounds).

Year	1221 1232	1242	1248	1254	1260 1262 1268	TOTAL
1957		17.6	18.7	23.3	31.9	91.5
1958	0.7	10.9	13.2	19.8	26.4	71.0
1959	1.1	14.3	16.7	23.1	30.8	86.0
1960	0.7	17.2	18.7	26.4	34.1	97.0
1961	0.9	18.7	23.1	29.0	37.4	109.1
1962	0.9	19.8	24.2	29.7	38.3	112.9
1963	1.0	17.6	23.1	29.0	38.5	109.2
1964	1.0	23.3	27.7	35.4	44.0	131.5
1965	0	29.7	37.4	44.0	49.5	160.6
1966	0.7	39.4	44.2	50.6	58.3	193.2
1967	0.9	41.8	48.4	53.9	61.6	206.6
1968	0	44.2	49.7	59.4	63.8	217.1
1969	1.3	46.2	50.6	61.6	67.1	226.8
1970	2.2	49.5	53.9	68.2	72.6	246.4
TOTAL	11.3	390.2	449.7	553.5	654.3	2058.9

Erickson (1997) estimates Monsanto's total production through 1976 at 1250 x 10^6 pounds, so there is either some error in the values or in my estimations. The Aroclor 1242 production figures are quite close to those provided by DeVoogt and Brinkman (**Table 1.3**); therefore, it may be considered likely that these included the 45 million pounds sold to NCR between 1959 and 1971 for carbonless copy paper. On the other hand, the nearly 10-fold higher production of Aroclor 1248 in the

Hutzinger estimates may indicate an entry error and correcting for the 400 million pound excess of Aroclor 1248 would bring the 2059 million pounds closer to other estimates (1.6 billion pounds vs 1.3 billion pounds).

Table 1.3. Domestic sales of Monsanto's PCBs in the US as presented by DeVoogt and Brinkman (1989) (10^6 pounds).

Year	1242	1248	1254	1260	Other	Total
1957	18.2	1.8	4.5	7.6	0.3	32.4
1958	10.4	2.6	6.7	6.0	0.4	26.1
1959	13.6	3.4	6.8	6.6	1.0	31.4
1960	18.2	2.8	6.1	7.3	0.8	35.2
1961	19.8	4.0	6.3	6.5	0.9	37.5
1962	20.7	3.5	6.3	6.6	1.0	38.1
1963	18.5	5.0	5.9	7.6	1.1	38.1
1964	23.6	5.2	6.3	8.5	1.2	44.8
1965	31.5	5.6	7.7	5.8	1.1	51.7
1966	39.6	5.0	7.0	5.9	1.6	59.1
1967	43.1	4.7	6.7	6.4	1.6	62.5
1968	44.9	4.9	8.9	5.3	1.2	65.2
1969	45.5	5.7	9.8	4.4	1.8	67.2
1970	48.6	4.1	12.4	4.9	3.1	73.1
TOTAL	396.2	58.3	101.4	89.4	17.1	662.4

Now that the Aroclor compositions are reasonably established (Frame et al., 1996a, 1996b), accurate production figures would enhance our ability to estimate the proportions of different congeners released into the environment. The above exercise in comparisons will NOT permit totally accurate estimations, but will be used in **Chapter 2** to derive approximate proportions of congeners produced.

PCBs were manufactured by the "random" chlorination of biphenyl until the desired properties, mainly viscosity, were achieved (**Table 1.4**). Individual chlorobiphenyls (CBs) (**Chapter 2**) are mostly solids, but the mutual depression of melting points within the mixture was responsible for the physical state: the more lightly chlorinated products were liquid while the heaviest chlorinated products were resinous or waxy. During aqueous solvation and/or volatilization, the more lightly chlorinated CBs are lost most rapidly. This also influences partitioning among compartments in the environment (**Chapters 4 and 5**) as well as within organisms (**Chapter 7**).

Table 1.4. Some physical properties of Aroclor mixtures (Hutzinger et al., 1974).

Aroclor	State	Density @ 20°C	Viscosity @ 99°C	Vapor Loss 6 hr @ 100°C	Water Solubility µg/L @ 25°C
1221	Oil	1.18	30-31	1-1.5 %	
1232	Oil	1.27	31-32	1-1.5 %	
1016	Oil	1.37			420
1242	Oil	1.38	34-35	0-0.4	240
1248	Oil	1.44	36-37	0-0.3	52
1254	Viscous	1.54	44-48	0-0.2	12
1260	Resin	1.62	72-78	0-0.1	3

a) Vaporization and aqueous solvation are selective, the lower chlorinated congeners being favored and lost first.

There are 10 vacant positions on the biphenyl rings (**Figure 1.1**), permitting 10 homolog groups and 209 possible substitution patterns (congeners) **(Appendix Table 1)**. Congener nomenclature has experienced various shorthand forms and attempts at standardization (Erickson, 1997, pp 20-22). The most ambitious and generally accepted is that by Ballschmiter and Zell (1980, referred to as "B&Z" or "BZ") as modified by Schulte and Malisch (1983) and Ballschmiter et al. (1987; 1992). The BZ nomenclature differs from actual IUPAC numbering in some instances (Guitart et al. 1993; Frame et al., 1996a), but in most instances the modified BZ numbers are retained to avoid confusion. The more common variations are noted in the tabular presentation of complex lists.

The separation of congeners for analysis is accomplished by gas chromatography (GC) which has enjoyed significant progress in the last 2 decades (Erickson, 1997; Frame et al., 1996b). The most sensitive detection has been with the electron capture detector which is quite sensitive to halogens. Identification is usually by retention time on a given separation system and confirmation (or less sensitive detection) is by mass spectrometry (**Chapter 3**).

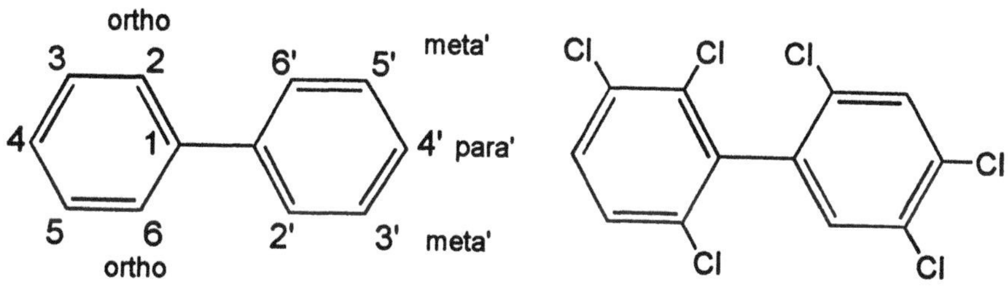

Figure 1.1. Structure of biphenyl and positions of substitution for CB 149. Note that 2,2',3,4',5',6-hexaCB could also be designated 236-245 or (incorrectly) 256-346.

The numbers of isomers per homolog group are presented in **Table 1.5**. Principles of chemical substitution favored certain vacant positions and patterns over others and the congener composition of most products was reasonably constant. Nevertheless, **reported compositions** varied as analytical instrumentation and techniques improved and increasing numbers of congeners became available for standards (Ballschmiter et al., 1992; Erickson, 1997; Frame et al., 1996a; 1996b).

Table 1.5. Characteristics of PCB homolog groups (Ballschmiter et al., 1992).

No. Chlorines	Mean Molecular Weight	% Chlorine	No. of Isomers
0	154.2	0	1
1	188.7	10	3
2	223.1	32	12
3	257.6	41	24
4	292.0	49	42
5	326.4	54	46
6	360.9	59	42
7	395.3	63	24
8	429.8	66	12
9	464.2	69	3
10	498.7	71	1

Caution is warranted in being overly dependent on earlier reports of PCB compositions. Only recently has the monumental task of resolving, identifying and quantitating all 209 congeners in the different Aroclors and in different lots of the same Aroclors been satisfactorily accomplished (Frame et al., 1996a).

The most recent reports of the distribution of homologs in different Aroclor products is presented in **Table 1.6** and congener compositions are presented in **Appendix Table 2**. The reader is again reminded that several alternative, although less correct, nomenclatures are possible.

Table 1.6. Approximate homolog distribution in different Aroclors (Frame et al., 1996a; 1996b).

No. Cl	A-1232	A-1016	A-1242	A-1248	A-1254	A-1260
1	31.3	< 1	< 1	< 0.2		
2	23.7	21.2	14.7	< 1	< 0.1	
3	23.4	51.5	46.0	20.9	1.8	< 0.3
4	15.7	27.3	30.6	60.3	17.1	< 0.3
5	5.8	< 0.6	8.7	18.1	49.3	9.2
6			< 0.3	0.8	27.8	46.9
7				< 0.3	3.9	36.9
8					< 0.05	6.3
9					< 0.05	0.7

Although some lots of Aroclors 1242 and 1248 differed modestly in composition, there were definitely at least 2 distinct types of Aroclor 1254 (**Appendix Table 2**). It is believed that the lot(s) with greater proportions of mono-*ortho* congeners 56, 66, 70, 74, 105, especially 118, and 156 arose from further chlorination of the residual mixture remaining after distillation of tetraCBs to produce Aroclor 1016 (JF Brown, Jr. personal communication, March, 1997). Therefore, only lots of Aroclor 1254 produced after 1970 would resemble this composition which is more like 2,3,7,8-tetrachloro-*p*-dioxin (TCDD; see **Chapter 2**) in toxic properties than earlier mixtures.

This overview of the composition and properties of former commercial PCB mixtures was purposely brief and contained information with which most experts are thoroughly familiar. PCBs in the environment, however, do not closely resemble a single commercial mixture and it is necessary to consider properties of individual congeners (**Chapter 2**) in order to predict the compositions of environmental mixtures. Analytical considerations are critical to gaining an accurate concept of PCB residue composition; there are several excellent recent reviews so the topic is merely introduced in **Chapter 3**. Physical and chemical properties interacting with various forces will mostly determine environmental disposition (**Chapters 4 and 5**). Discussions of biotic metabolism and disposition and actual residues are adequately broad and complex to require most of the remainder of the book.

2 THE INDIVIDUAL CHLOROBIPHENYLS

The 209 individual chlorobiphenyls (CBs) have been listed in **Appendix Table 1** by structures and B & Z designations which approximate IUPAC numbers. The average percent compositions of important Aroclors are presented in **Appendix Table 2**. Congener proportions in individual Aroclors are discussed in more detail in **Section 2.3**. Sample properties of the 209 congeners are presented in **Appendix Table 3**. Estimates of relative proportions of congeners produced as well as additional Aroclor compositions are given in **Appendix Table 4**.

It is quite important to recognize that the physical-chemical "constants" presented are averages or selected values from multiple sources. This chapter emphasizes trends rather than absolute values. The constants presented are reasonable estimates, but solubilities, partition coefficients, and volatilities may differ by at least an order of magnitude between studies. Half-lives and bioconcentration factors, discussed in later chapters, are often even less directly comparable depending on the laboratory and the methods. For comparative purposes, individual properties such as volatilities, solubilities, toxicokinetic parameters, etc. might be useful from single sources; however, most studies develop values for only a portion of the 209 CBs and completing the comparisons by merging data from multiple authors must be conducted with caution. This is especially true for half-lives and bioconcentration factors. Mackay et al. (1992) have compiled a large number of properties from several authors for many of the CBs. This is an extremely useful reference for selecting and comparing physical-chemical constants. It also emphasizes the ranges of values from which one must select in order to have "constants" for developing models and relationships among various environmental chemicals, including PCBs.

2.1 GC Retention Times

Properties of PCB congeners vary in a systematic manner. Trends can be readily discerned with increasing degree of chlorination, but the systematic variations in physical and chemical parameters are not linear with increasing B & Z number. Gas chromatography and retention times are discussed in **Chapter 3** (3.1). The relative GC retention times under specified conditions are probably the most readily determined constants and structure-retention relationships are quite useful for developing concepts of other relationships. The generally increasing, but regular variations, of relative retention times with B & Z number were plotted (Mullen et al., 1984) and this plot has been reproduced (Erickson, 1997).

The chlorine substitutions responsible for irregular progressions of properties with degree of chlorination can also be seen with more limited data sets. Retention times of symmetrical and 2,4,5-chlorinated congeners only are presented in **Table 2.1** and plotted in **Figure 2.1** (Data from Cochran and Frame, 1999). Melting points would plot even more irregularly (**Appendix Table 1**), with the rare 2,4,6-chlorinated congeners (e.g., 50, 69, 75, 100, 115) dropping well below neighboring IUPAC numbers. The also rare non-*ortho* (3,4 and 3,5) CBs generally have higher melting points as well as longer RTs.

Table 2.1. Retention times of symmetrical CBs and those containing 2,4,5-chlorination on one ring as a function of B&Z number (methyl-polysiloxane column) (Cochran and Frame, 1999).

B&Z No.	Chlorine Position	Ret Time (min)	B&Z No.	Chlorine Position	Ret Time (min)
4	2-2	22.15	133	235-235	56.37
11	3-3	28.70	136	236-236	51.33
15	4-4	29.67	138	234-245	60.60
29	245	32.97	146	235-245	57.08
40	23-23	42.13	149	236-245	54.77
47	24-24	38.73	153	245-245	57.80
48	245-2	38.92	154	245-246	52.47
52	25-25	38.00	155	246-246	47.03
54	26-26	32.53	167	245-345	64.10
74	245-4	44.17	169	345-345	70.93
77	34-34	51.47	180	2345-245	68.85
80	35-35	45.18	183	2346-245	63.58
97	245-23	48.40	187	2356-245	63.02
99	245-24	50.07	194	2345-2345	79.35
101	245-25	47.88	197	2346-2346	68.53
102	245-26	44.88	202	2356-2356	66.90
118	245-34	54.85	203	23456-245	73.93
120	245-35	51.45			
128	234-234	63.35	209	23456-23456	87.30

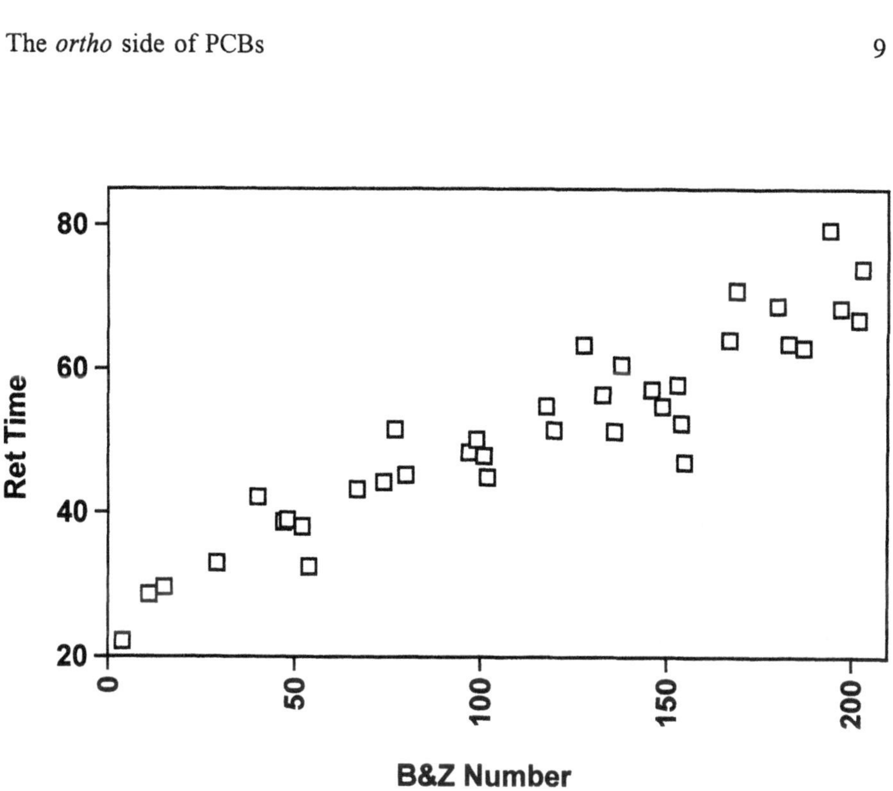

Figure 2.1. Graphic representation of data in **Table 2.1**.

Indistinguishable retention times (poorly resolved CBs) also indicate similar physical properties. Poor resolution introduces analytical and reporting complexities which may result in abridged or confusing reports. These reports, then, may have a great impact on forming accurate perceptions of residue as well as, e.g., Aroclor compositions. The environmentally important CBs 95, 132, 149, and 163 are frequently overlooked if only the "major" CB of the peak is included in analytical reports or if thorough analytical reports are abridged. **Table 2.2** reminds the reader of important co-eluting congeners, but also helps to illustrate that *ortho* chlorines may accelerate elution for heavier congeners while *para* chlorines tend to retard movement of even lighter CBs through GC columns.

Table 2.2. Comparison of chlorine substitution patterns of some co-eluting non-homologous chlorobiphenyl pairs.

CB Pair	Less Chlorinated			More Chlorinated		
	Sum Cl	*para*	*ortho*	Sum Cl	*para*	*ortho*
15/17	2	2	0	3	1	2
33/53	3	1	1	4	0	3
66/95	4	2	1	5	0	3
77/110	4	2	0	5	1	2
118/149	5	2	1	6	1	3
105/132	5	2	1	6	1	3
105/153				6	2	2

A detailed discussion of Structure-Retention Relationships on various GC columns is presented by Cochran and Frame (1999).

2.2. Other Properties

This chapter does not intend to delineate properties (which can be derived from the references), but rather to suggest trends related to *ortho*-chlorination. The scarcity of *ortho* chlorines has been emphasized in the literature for the past decade because non-*ortho* and mono-*ortho* chlorinated congeners with lateral chlorines can assume the coplanar configuration required for binding to the aryl hydrocarbon receptor (*Ah*R). Such binding initiates a specific spectrum of biological responses (**Section 2.4**). Homologs which are *ortho*-poor would, by necessity, be more *meta-*, *para-*rich and vice versa. The focus of this book is the more prevalent CBs which are *ortho*-rich, but neither can be considered in a vacuum. Properties such as GC retention times and octanol:water partition coefficients tend to increase with increasing chlorination, but decrease within a homolog group as the congeners become relatively rich in *ortho* chlorines. Conversely, water solubilities and vapor pressures decrease with increasing chlorination, but rebound within homolog groups with increasing proportions of *ortho*-chlorination. The homolog numbering system (lowest digit first) has a predictable impact on the sequence of chlorination patterns. The first CBs in a homolog group are *ortho, meta* rich (2,3); the 6 and 6' *ortho* chlorines are on higher-numbered homologs and the non-*ortho* (3,4,5) congeners have the highest numbers.

2.2.1. Volatility

Lightly chlorinated and *ortho*-rich congeners are more volatile (as might be predicted from GC retention times) and tend to be more water soluble. These properties render them more mobile in the environment and, indeed, atmospheric and water profiles are enriched in lightly chlorinated and *ortho*-chlorinated congeners compared to profiles in other media (**Chapter 4**).

Vapor pressures are strongly dependent on temperature and whether the chemical is in the solid or liquid phase. Considerable uncertainty can arise from determinations for chemicals of low volatility such as most PCBs (Mackay et al., 1992). If selected congeners are compared according to best-estimate vapor pressures, it appears that *meta* chlorines may decrease volatility from the solid phase more than *para* chlorines once tetra-chlorination is achieved (**Table 2.3**).

Table 2.3. Vapor pressures for selected CBs in both solid and liquid phases and Henry's Law Constants at 25°C (Mackay et al., 1992).

B&Z No.	Chlorines	Vapor Pressure (Pa)		Henry's Law Constant (Pa/Sol)
		Solid	Liquid	
4	2-2	0.265	0.6	59.17
11	3-3	0.027	0.03	17.26
15	4-4	0.0048	0.08	17.00
18	25-2	0.143	0.22	92.21
30	246	0.0384	0.09	49.51
33	2-34	0.0136	0.003	43.67
40	23-23	0.00225	0.002	21.94
47	24-24	0.0054	0.002	17.38
52	25-25	0.0049	0.002	47.59
77	34-34	0.000059	0.002	1.72
87	234-25	0.000304	0.0023	24.81
101	245-25	0.00109	0.0035	35.48
128	234-234	0.00002	0.00034	11.91
153	245-245	0.000119	0.0007	42.9
TCDF	2,3,7,8	0.000002	0.0002	1.5
TCDD	2,3,7,8	0.0000002	0.0001	3.3

Volatility and solubility influence fugacity or escaping tendency (**Chapter 5**) and both govern Henry's Law Constant which "is essentially an air-water partition coefficient" (Mackay et al., 1992). Low volatilities and low solubilities again make determinations for PCBs somewhat difficult. Values calculated from vapor pressure and solubility are included in **Table 2.3** and negative log values of Henry's Law constants for all 209 congeners are included in **Appendix Table 3**. The contributions of both volatility and solubility cause Henry's Law Constants for PCBs to vary in patterns somewhat different from other properties. Overall, however, water surface layers and atmospheres are enriched in lightly chlorinated and *ortho*-rich congeners and more specific information is provided in **Chapter 4** and **Chapter 5**.

2.2.2 Partition Coefficients

Water solubilities contribute to both air:water partitioning and aqueous:lipid partitioning. More lightly chlorinated and *ortho*-rich congeners tend to be more water-soluble. As with other "constants", care must be exercised in deriving octanol:water partition coefficients which are critical for comparative and modeling purposes. Several Log P values are included in **Appendix Table 3**, both as selected or averaged values from Mackay et al. (1992) and single values from Koslowski et al. (1994). Although there is reasonably close agreement, ranges of Log P values from Mackay et al. are presented in **Table 2.4** as a further caution against filling in missing values from several different studies. It appears to this author that the selected Log P for CB 18 is too high relative to the other CBs.

Table 2.4. Ranges and selected Log K_{ow} values of chlorobiphenyls (Mackay et al., 1992).

B&Z No.	Log K_{ow}	Range
15	5.3	4.8 - 5.6
18	5.6	3.9 - 6.2
33	5.8	5.5 - 6.1
40	5.6	4.5 - 6.3
47	5.9	5.2 - 7.1
52	6.1	3.9 - 6.9
77	6.5	5.3 - 6.6
128	7.0	6.1 - 8.3
153	6.9	4.5 - 8.4

2.2.3. Atropisomers

One consequence of *ortho*-chlorination of the biphenyl molecule is the hindrance of ring rotation by the bulky chlorines. Two *ortho*-chlorines prevent the rings from assuming a coplanar orientation, but three or four *ortho*-chlorines can lock the two rings into chiral stereoisomers. Nineteen have been predicted to exist as stable enantiomers at ambient or physiological temperatures (**Table 2.5**).

Table 2.5. CBs forming stable atropisomers. The most environmentally important congener structures are in bold type.

2,3,6-Pattern	Chlorine Position	2,3,4,6-Pattern	Chlorine Position
45	**236-2**	88	2346-2
84	**236-23**	131	2346-23
91	236-24	139	2346-24
95	**236-25**	144	2346-25
132	**234-236**	171	2346-234
135	235-236	175	2346-235
136	236-236	176	2346-236
149	**236-245**	183	**2346-245**
174	**2345-236**	196	2345-2346
176	2346-236	197	2346-2346

The racemic mixtures can be formed during manufacture and both optical isomers of several congeners have been detected in various biota, including human breast milk (e.g. Glausch et al., 1995; Reich et al., 1998; Cochran and Frame, 1999). It has also been demonstrated that the chiral isomers have different biological potencies and different susceptibilities to biodegradation enzymes (Putman et al., 1989; Reich et al., 1998; Rodman et al., 1991).

Atropisomers of PCB methyl sulfones have also recently been shown to be selectively retained in human liver (Bergman et al., 1998).

2.2.4. Energies of Formation

The energies of formation for various halogenated aromatics has been a topic of interest because of the theoretical importance to microbial dechlorination processes (e.g. Holmes et al., 1993). Since several parameters must be considered in the estimations of Gibbs free energies of formations for PCBs, these exercises also provide good sources of consistently derived constants for the full spectrum of congeners. Gibbs free energies of formation as calculated are included in **Appendix Table 3.**

Energies of formation have also been proposed as predictors of toxicities (Genoni, 1997), but this is acknowledged to be only an idea which may supplement other quantitative structure:activity relationships (QSARs) (see Schultz, 1997). Energies of formation should also influence the proportional yields of congeners resulting from random .chlorination of the biphenyl molecule. We attempted such a correlation out of curiosity (LG Hansen and MA Khan, unpublished results) and there did appear to be a reasonable relationship.

2.3. Congener Abundance in Former Commercial Mixtures.

For profile comparisons, it should also be useful to estimate the relative yield of each of the congeners over the range (1-9) of chlorine additions. To derive a simple estimate, the most recent compositions reported for the major Aroclors produced by truly random chlorination were averaged from Frame et al. (1996). The Aroclor 1016 and residue-chlorinated Aroclor 1254 were excluded because they did not represent random mixtures.

The derivation of relative congener yields above led to an estimation of the relative amounts of each congener produced by the random chlorination. The "yield" of each congener in a specific Aroclor was multiplied by a "production" factor derived from **Tables 1.2** and **1.3** which list reported domestic sales of Aroclors in the USA. Even if the actual sales figures deviate (which is likely because of the differences between the two reports for the same time period), the ratios of Aroclor sales are probably reasonable. For simplicity, it is assumed that USA sales ratios reflected global usage patterns and that the compositions of other PCB mixtures were similar to the corresponding Aroclor mixture. Estimated production ratios from the two reports which were used in the calculations are presented in **Appendix Table 4**. The largest discrepancies were in the relative sales of Aroclors 1242 and 1248, which have reasonably similar profiles.

The resulting relative productions of congeners for the two sales estimates are presented in **Appendix Table 4**. This simple exercise has probably been conducted with more accurate production figures, but the outcome is to emphasize the congeners which were clearly produced in the highest amounts (**Table 2.6**). Of course, in the more recent (1971-1977) production, Aroclor 1016 became an important contributor and the high TEQ Aroclor 1254 would distort the profiles somewhat (**Appendix Table 2**).

Table 2.6. Relative proportions (percent total) of major CBs produced based on Aroclor compositions and Aroclor sales in the USA (from **Appendix Table 4**).

B&Z No.	From Tab 1.3	From Tab 1.2	B&Z No.	From Tab 1.3	From Tab 1.2
4	1.96	0.70	87	1.06	1.53
8	4.47	1.63	95	1.82	2.87
16	1.99	0.84	99	0.89	1.24
17	1.99	0.86	101	2.26	3.63
18	5.53	2.56	105	0.91	1.26
22	1.84	0.86	110	2.35	3.59
28	4.59	2.36	118	1.80	2.69
31	4.95	2.62	126	0.001	0.001
32	1.24	0.58	128	0.31	0.56
33	3.25	1.51	132	0.79	1.53
44	3.01	2.48	138	1.89	3.64
49	2.07	1.63	149	1.86	3.76
52	3.54	3.46	153	1.98	4.00
56	1.46	1.15	156	0.20	0.39
66	2.79	2.26	163	0.51	1.04
70	3.45	3.16	180	1.80	3.83
74	1.58	1.38	187	0.87	1.83
77	0.23	0.16	194	0.32	0.69
81	0.008	0.006	203	0.24	0.49

Radically asymmetric substitutions are rare (e.g. CBs 86, 93, 108, 112, 142, 160, 166, 181, 186, 192), although some such as CBs 81, 114, 173, 185 and 190 do occur in small amounts. Some substitution patterns such as 3,5 and 2,4,6 are known to be rare while 2,5 and 2,4,5 substitutions are the most common; this is confirmed in **Appendix Table 4**. Examples of rare 3,5-chlorinated congeners include CBs 80, 107, 111, 113, 120, 121, and 127 while the rare 2,4,6 pattern includes CBs 98, 100, 103, 104, 119, 148, 150 and 155. Many studies used mainly the rare congeners and/or their metabolites which made them vulnerable to criticism for lack of relevance.

The congeners which persist and bioaccumulate are readily apparent from the legions of residue studies conducted; major residue CBs 74, 99, 128 and 194 were produced at relatively lower levels than other CBs, attesting to their persistence. The congeners *produced* are important because of sporadic releases of relatively unchanged former commercial mixtures. Even though the lighter Aroclors are generally considered minor products, the very high proportions of mono- and di-CBs can account for the sporadic detection of these labile congeners, especially CBs 4 and 8. The high-production/low-residue congeners are not only candidates for sporadic exposures, but must certainly contribute a high proportion of metabolites to the global burden.

Of the important frequently co-eluting congener pairs, it may be worth noting the relative proportions produced since these are generally not specified: 16>32, 31=28, 56>60, 70>>76, and, of course, 110>>>77 (**Table 2.6**; **Appendix Table 4**). It has been known that CB 138 is about 3 times higher than CB 163, but the nearly equal production of CBs 66/95, 105/132 and 118/149 is frequently overlooked. The separate consideration of co-eluting (and, thus, somewhat similar) congeners is important only if persistence and biological properties are different.

2.4. Congener-Specific Biological Properties.

Co-eluting CBs 77 and 110 are both relatively rapidly metabolized, but CB 77 is a representative coplanar AhR agonist while CB 110 is a potent PB-type inducer (Li et al., 1998). CB 28 is known to be much more persistent than co-eluting CB 31. Co-eluting CBs 66/95 are known to have highly divergent potencies at ryanodine-sensitive calcium channels (Wong and Pessah, 1996). Similar relationships between other mono-*ortho* and multi-*ortho* pairs are emerging (Pessah et al., 1999).

Several compilations and reviews of PCB congener toxicities have been published in the past 2 decades. The actions and potencies of coplanar AhR agonists are established and well-documented (Safe, 1994; Ahlborg et al., 1994). Lightly chlorinated and multi-*ortho* congeners have generally been collectively referred to as "other", "di-*ortho*", and/or "inactive." Consistent with the physical-chemical properties, however, there are unique patterns of biological activities among the lightly chlorinated and *ortho*-rich CBs. DiCBs 4 and 15 and triCBs 18 and 28 are examples of totally contrasting pairs of lightly chlorinated homologs (See Hansen, 1998).

For the *ortho*-rich tetra- through heptaCBs which are environmentally relevant, there are at least 3 "types" based on endocrine, enzyme, neurochemical and vitamin depletion responses. Enzyme (especially CYP P450s) induction currently provides the most complete link between structures and biological actions, but SARs for this well-studied relationship are still incomplete. Nearly all PCBs disrupt thyroid hormone homeostasis by one of several mechanisms, and transient hyper- as well as hypothyroidism can occur; the data for *ortho*-rich congeners are very scant and SARs are not possible at this time.

There has been a renewed interest in the nonplanar CBs and a variety of SARs are now emerging. However, adequate coverage of these aspects will require a separate monograph. It is hoped that:

1) Additional and more comprehensive SARs will be published in the near future, and
2) The following rather subjective table will provide a "work sheet" for developing additional data on the congeners which this monograph will demonstrate to be significant contributors to exposure profiles.

Table 2.7. Examples of molecular, biochemical, morphological and functional effects of chlorinated aromatics mediated through the Ah Receptor, mediated by TCDD-like compounds independent of the Ah Receptor, and mediated through mechanisms distinct from that of TCDD-like compounds (adapted from Hansen, 1997; 1998b). Relative increase (+) or decrease (-); no effect = 0; (probable).

Examples of Effects	TCDD TEQ AhR	TCDD Non-AhR	Non-TCDD PCB
ENHANCED GENE EXPRESSIONS			
Protooncogenes	+	+	+
Interleukins	+	?	
TGFs	+	?	
EGF Receptor	+	?	
CYP 1A1, 1A2, 1B1	++	(+),(-)	0
CYP 2A, 2B, 3A	**0(-)**	?	++
UDP-Glucuronyl Transferases	++		+
Glutathione Transferases	+		+
Aldehyde dehydrogenase	+		
GENERALIZED METABOLISM			
Steroid synthesis	-	-	?

Examples of Effects	TCDD TEQ AhR	TCDD Non-AhR	Non-TCDD PCB
Steroid catabolism	+		+
Adrenal	+&-		+&-
Vitamins A,D,E	-		-&+
Vitamin K; coagulation	+/-		--
OTHER HEPATIC			
Liver Hyperplasia	+		
Bile duct hyperplasia	+		+
Liver hypertrophy	+	+	+
Tumor Promotion	+		+
Fatty Liver	++	(+)	
Porphyria	++	+	+
GASTROINTESTINAL			
Urinary bladder hyperplasia	+		
Intestinal hyperplasia	+		
Digestive tract inflammation	+		+
Ulcers	+		+
Digestive tract cancers	+		
HEMATOPOIETIC			
Anemia	+	+	+
Neutrophilia	?		+
Neutrophil activation	0		+
REPRODUCTIVE			
Embryo Lethality	+++	+	++
Teratogenesis	++		+/-
Ureter ---> hydronephrosis	++		
Fertility (female)	?		-
Endometriosis; cancer	++	?	(+)
Fertility, Libido (male)	-		?

Examples of Effects	TCDD TEQ AhR	TCDD Non-AhR	Non-TCDD PCB
COMPLEX DEVELOPMENTAL			
Thyroid homeostasis	--	-	--&+
Retinoid homeostasis	--	?	-
Estrogenic	?	+&-	+
Antiestrogenic	++	(+)	+&-
NEUROCHEMICAL BEHAVIORAL			
Dopamine increase	+		?
Dopamine decrease	0		+
Calcium Channel Activation	0		+
Neuron 2nd messengers	0		+
Avoidance Responses	?		-
Spatial Learning	+		-
Activity	-, 0, +		- & ++
OTHER			
Wasting syndrome	+		0
Body weight	--	(-)	-
Insulin release	0		+
Intercellular communication	0		-
Skin hyperplasia-chloracne	+		
Urinary bladder hyperplasia	+		
Thymic atrophy	--		0
Other immune (spleen, etc)	--	?	-
Oxidative Stress	++	+	+

3 INTERPRETATION OF ANALYTICAL DATA

PCB chemistry is relatively straightforward (Chapter 2), but the possible presence of 209 CBs along with metabolites and atropisomers makes some selection necessary for comprehension. More recent efforts can focus on fewer congeners (**Table 2.6**; **Appendix Table 4**), but the data are still extensive. Even when restricted to the 60 or so most prevalent congeners and the highly potent coplanar congeners, the continuum of data collection→presentation→comprehension can be overwhelming. The large ranges in partition coefficients, volatilities, and bioconcentration factors insure that the mixtures will have different profiles in different matrices. This adds another level of complexity to the accurate assessment of PCB exposures.

Even scanning an analytical paper can be challenging for the non-analyst. Relatively straight-forward PCB residue reports also have incremental degrees of difficulty depending on the level of information required by the reader. Highly abridged reports or those purposely limited in scope are the most popular because of the practically indigestible data generated by more comprehensive PCB analyses. Unfortunately, this leads to misconceptions regarding residue profiles and exposures which then encourages reporting of still fewer congeners. The overly narrow understandings may also influence risk assessment and research directions which can, in turn, perpetuate abridgement of residue reports (Hansen, 1998). For general unabridged concepts, it is still better to use the equivalent of packed-column peaks or categories than to selectively exclude analytes from a high resolution analysis.

The objective of this chapter is not to discuss analytical methodologies. These are adequately addressed by professional analysts in recent works which also present thorough critiques of earlier methods (e.g. Cochran and Frame, 1999; Erickson, 1997; Frame et al., 1996b; Hess et al., 1995). Likewise, judgments regarding quantitative accuracy and precision are deferred to the analysts; however, qualitative accuracy (i.e. CBs probably present vs those reported) is important to the overall theme of this monograph.

The intent is 1) to highlight details necessary for accurate analytical interpretations, 2) to compile and compare analytical results to help establish a more comprehensive concept of PCB residues, and 3) to suggest methods for both presentation and digestion of these extensive data. If successful, those of us who are not analysts can begin to comprehend and profit from the extensive amount of information available.

3.1. Low Resolution GC and Pattern Recognition

A general description of gas chromatography could be paraphrased, but Erickson (1997, Chapter 7) provides a succinct overview useful for the novice:

> "GC, also called gas-liquid chromatography, has been a workhorse technique in most analytical laboratories for over 25 years. The function of a GC is to separate complex mixtures and detect the components. The instrumentation consists of a pressurized gas (generally purified nitrogen, helium, or hydrogen) source, an injector, the separation column, and a detector. A very small (generally 1 to 10 μL) amount of sample is injected at the head of the column with a microsyringe, where it is vaporized. The vapors are swept by the carrier gas onto the column, where the components are separated, and then to the detector, where they are detected. *The detection output consists of a series of peaks with time.* (emphasis added) The intensity of the peaks is generally proportional to the amount of compound present. The separation is effected by the interaction of the compounds with the gas and liquid phases. The more soluble the compound in the liquid phase, the more time it will spend dissolved in the liquid phase and the later it will elute. In addition, GC analyses use elevated temperatures or a programmed-temperature range to enhance separation, so compound volatility also affects the retention time. PCBs generally elute in order of chlorination, $C_{12}H_9Cl$ first and $C_{12}Cl_{10}$ last; although there is considerable overlap in the middle homologs....
>
> The separated peaks exit ("elute") from the GC column and are detected at different times. Measurement of the time required to elute through a given column yields a retention time (RT) which is reproducible for a given compound. RTs are often measured relative to an internal standard to compensate for fluctuations in temperature or carrier gas flow rate. The relative retention time, RRT, can be used to identify components by comparison to RRTs of an authentic standard. Selective detectors provide an added dimension to the identification of PCBs...."

RRTs for PCB congeners are discussed in detail in **Section 2.1**.

Early GC methods for PCB analysis were generally based on packed column separations and highly sensitive electron capture detection. More extended analytical techniques including better resolution and mass spectroscopy provided the basis for application of these more generic techniques to PCB estimation. Detector responses to peaks (containing multiple CBs) were compared to detector responses to peaks of the same RRT in commercial mixtures, usually Aroclors. Using extensions of the existing pesticide methods, "typical Aroclor pattern(s)" were readily discernable (See Chapter 8 in Erickson, 1997). Once the Aroclor pattern was determined, 3 to 5 marker peaks may have been used to estimate the equivalent concentration of that particular Aroclor required to cause the same detector response for these peaks.

Erickson (1997) lists and discusses the US EPA Method 8080 series of PCB-as-Aroclor directives in his Chapter 3 and Chapter 8. The modest guidance provided is quoted as:

"PCB residues are quantitated by comparison to one or more of the Aroclor materials, depending on the chromatographic pattern of the residue. A choice must be made as to which Aroclor or mixture of Aroclors will produce a chromatogram most similar to that of the residue. This may also involve a judgment about what proportions of the different Aroclors to combine to produce the appropriate reference material."

Relatively unchanged PCB mixtures (such as transformer oils) may be adequately quantitated by pattern recognition techniques. This method, however, is unsuited for environmental samples more removed from the original mixture. Analysts have long known that these profiles only vaguely resemble the original mixtures and force-fits introduce serious quantitative errors. With appropriate attention to all peaks, more accurate total PCB and changes in residue profiles can be detected (e.g, Boon and Duinker, 1986; Dieter, 1974; Krauthacker et al., 1986; Yakushiji et al., 1979). However, long-term trends must reconcile packed column quantitations with specific-congener methods. Some series may be comparable, but applying newer methods to older samples revealed a trend opposite to that initially perceived for Lake Ontario sediments (Oliver et al., 1989). Erickson (p 344) also cites an example where PCB temporal trends in bird eggs were reversed by using a more accurate quantitation method.

3.1.1. Pattern Characterization

Early efforts at PCB analyses were also hampered by a lack of identified congeners within the mixtures. For this reason, the cruder resolution of packed column GC was probably an advantage. Collectively, there were fewer "unknowns" and the data were more consistent. Different peaks were seen to behave in characteristic manners even though it was known that each contained a variety of different CBs. The efforts of pioneers (e.g. Sissons and Welti, 1971; Webb and McCall, 1972; 1973; Jensen and Sundstrom, 1974; Kuroki and Masuda, 1977) permitted characterization of the major components of each peak in the commercial mixtures and helped to establish which CBs persisted in biota. This made peak accumulation/disappearance more meaningful since at least some structures could be assigned. It also allowed selection of probable environmentally significant congeners for biological testing.

For direct dosing studies, comparison of the residue profile with the administered PCB mixture revealed distinctions among the various peaks (e.g. Grant et al., 1971; Fries et al., 1973; Bush et al., 1974; 1975; Borchard et al., 1974; 1975; Hansen et al., 1975; 1977). These differences could be quantitated by using the dose mixture as the standard and converting residues to relative amounts of standard (Aroclor) required to elicit the same peak response. In most of these studies, the relative contributions of all major peaks were determined separately rather than merely using selected peaks to estimate total Aroclor equivalents.

Relatively consistent Aroclor patterns were obtained using *p,p'*-DDE as the retention time standard. Low resolution GC structures and peaks (Webb and McCall, 1972; 1973) as well as some interpretations of Aroclor and residue profiles are presented in **Table 3.1**. The dominant structures assigned to major peaks, however, were not consistent and led to some confusion. As might be expected, this confusion grew with the introduction of high resolution GC and the increased numbers of analytes detected (**Section 3.2**). Most of our work simplified peak contents and greatly overestimated the contributions of some congeners to some peaks while neglecting others. For example, we consistently suggested that peak 48 in Aroclor 1242 contained mainly CB 52 and accounted for about 25% of the composition of the Aroclor and we often mis-assigned peak 37 to CB 37.

A similar more ambitious comparison of high resolution profiles with Sissons and Welti (1971) and Jensen and Sundström (1974) identifications was developed by Bush and colleagues (1985a). These designations, with some corrections, were used to assess profiles in human milk (Bush et al., 1985b) and are included in **Appendix Table 5.**

Table 3.1. Representative chlorobiphenyls attributed to peaks of similar retention times relative to *p,p*-DDE in low resolution PCB analyses compared to peaks in evolving high resolution techniques.

Peak RRT[a]	B&Z # or (# Cls)	Presence in:		Peak Designation by:		
		A 1242[a]	A 1254[a]	Hansen[b]	Wolff[c]	Fait[d]
13	1	+				1
17	3	+				
18.5	4	+		21		2
23.5	6	+				7
24	8	+				5
29	18	+		26		11
29.2	17	+				12
29.5	15	+		31	1A	16
34	16	+				
38	25	+				

Peak RRT[a]	B&Z # or (# Cls)	Presence in:		Peak Designation by:		
		A 1242[a]	A 1254[a]	Hansen[b]	Wolff[c]	Fait[d]
	37			37x		
38.5	28	+		(37)	1	22
38.7	31	+		(37)	1	21
42.5	33	+		41		22
47	**(4)**		**6.2**			
48	52	+	+	48	2	24→25
49.5	49	+	+		3	25→27
50.5	47	+			4	27→28
54	**(4)**		**2.9**			
55	44	+	+	53	5	26
58	**(4)**		**1.4**			
59	96	+	+	57x		
70	**(4,5)**		**13.2**			
70	74				6	39
70	70	+		70	7	38
70.5	95		+	70	9	34
71	66	+		70	8	40
76	91		+			36
84	**(5)**		**17.3**			
83	84		+	83	14	36
85	101		+	85	11	45
87	99		+	85	12	47
98	**(5)**		**7.5**			
96			+			
99	97		+	99	16	48
	87				17	NR→55
101			+			

Peak RRT[a]	B&Z # or (# Cls)	Presence in: A 1242[a]	A 1254[a]	Peak Designation by: Hansen[b]	Wolff[c]	Fait[d]
104	**(5)**		**13.6**			
107	110		+	105	18	55→59
125	**(5,6)**		**15.0**			
125	149		+	127	21=146	54→55
126	118		+	127	20	60
146	**(5,6)**		**10.4**			63
147			+			
148	153		+	149	22	64
	105				23	63
149			+			
152			+			
160	**(6)**		**1.3**			
174	**(6)**		**8.4**			
	138			176	25	68
203	**(6)**		**1.8**	208		
232	**(7)**		**1**	253		

[a] Webb and McCall, 1972; 1973. Retention time relative to *p,p*-DDE = 100. Some peaks were characterized by mass only, but were included in the standards and structures suggested by others. Bold designations are low resolution peaks included in Aroclor 1254 standard and approximate percent of Aroclor 1254.
[b] Hansen et al., 1975; 1977; 1979; revised and updated in Hansen, 1987. Retention time relative to *p,p*-DDE = 100. An "x" indicates an incorrect designation.
[c] Wolff et al., 1982. Peak order number out of 37 separated by high resolution methods. More complete data in **Appendix Table 5**.
[d] Fait et al., 1989. Peak order number out of 97 separated by high resolution methods. Changes in designation (→) were provided by ED Pellizzari, personal communication, December, 1996. Complete data in **Appendix Table 5**.

Sample data for Webb and McCall peaks reveal that sources and species differences are readily discernable. A comparison of the relative contributions of these peaks to total PCB in food animals and in humans is presented in **Table 3.2**. Similar profiles can be found for sheep (Hansen et al., 1977) and catfish (Hansen et al., 1976a). Chicken and swine fat are also compared in terms of some high resolution analyses (Hansen et al., 1983). The total PCB estimations by this low resolution analysis are highly correlated with high resolution analysis of the same samples (Burse et al., 1994). This type of data permits formulation of general concepts without the overwhelming burden of complete congener profiles or the (unintentional) misrepresentation by abridged residue reports.

Table 3.2. Profiles (% total) of PCB peaks in blood following chronic Aroclor or multipathway exposures. Based on Webb and McCall designations (**Table 3.1**).

Peak RRT DDE = 1	Swine[a]		Chickens[b]		Human		
	1242	1254	1242	1254	Occupational[c]		Fish[d]
0.21	0.1						
0.26	0.8						
0.31	4.3						
0.37	15.0		25.0		35.9	29.2	10.3
0.41	4.1						
0.48	6.6		37.8		5.7	5.4	1.3
0.53	18.6		1.1		2.8	1.9	0.1
0.57	22.3		5.0		3	0.8	<0.1
0.70	17.4	4.1	15.8	9.3	21	31.3	17.9
0.78[e]					7.6	2.7	
0.84[e]	8.8	20	7.8	13.0	5.4	5.2	4.9
0.99	0.8	12.3	2.7	9.9			
1.05	0.6	4.6	1.5	6.3			
1.27	0.2	11.9	1.8	20.6	4.5	5.2	15.6
1.49	0.2	16.7	1.1	20.2	3.4	7.5	23.5
1.76	0.1	19.8	0.3	11.9	4.8	5.8	16.4
2.08		4.3		5.4	2.1	1.2	
2.53		4.0		2.4	1.8	1.0	9.9
2.86		2.0		0.8	1.9	2.5	

[a] Growing swine receiving 20 ppm in feed for 90 days (Hansen et al., 1977).
[b] Broiler cockerels receiving 20 ppm in feed for 56 days (Hansen et al., 1976b).

[c] Capacitor manufacturing workers in 1976 (first column) and 1979 (second column) (Wolff et al., 1982; 1992).
[d] Higher residue residents of the New Bedford, MA area who reported consuming fish and/or seafood on a regular basis (Burse et al., 1994). Only 13/of 23 individual profiles were used in this table; one deleted had an unusual profile (54% peak 37 or 78% CB 28 by specific congener analysis) and nine had unresolved contaminants which prevented quantitation of early eluting peaks.
[e] This series was variously resolved, but peaks 83 (persistent) and 85 (labile) were combined for simplicity.

3.2. Transition from Low Resolution to High Resolution GC

Although the comparisons to high resolution work may seem somewhat straightforward, the reader is invited to take the structural designations for the 37 peaks from Wolff et al. (1982) or, especially, the 97 peaks from Fait et al. (1989) and work out comparative profiles (without guidance from **Tables 3.1 or Appendix Table 5**). It is little wonder that inconsistencies and inaccuracies were introduced into the literature, especially since most of these data sets were constructed with theoretical retention times (complete congener standard sets were not yet generally available).

The author attempted to resolve certain discrepancies between the early designations by Sissons and Welti (1971) and Webb and McCall (1972; 1973) with more recent high resolution studies. Having gone through the effort (several weeks), some of the results are shared with the reader (**Appendix Table 5**). The published residues in occupationally exposed workers and other New York residents appear to be somewhat inconsistent (Erickson, 1997). Correspondence with analysts and deletion of structures generally shown to be absent from former commercial Aroclor mixtures (Frame et al., 1996a; see footnote to **Appendix Table 5**) simplified the data and improved the consistency.

Pellizzari's revisions in the Fait designations were very helpful. Bush also provided re-designations of some peaks such as BZ No. 82-to-110 and BZ No. 179-to-118 (B Bush, personal communication, January, 1997). Albro expanded his pioneering work with separations (Albro et al., 1977) to derive compositions of lower chlorinated (Albro and Parker, 1979) and higher chlorinated (Albro et al., 1981) Aroclors. As with the other studies, only a few standards were available and theoretical retention times of CBs 91 and 95 were too close to resolve; a best guess assigned the larger proportion of the Aroclor to CB 91, but this has since been revised (P Albro, personal communication, May, 1997; Frame et al., 1996a). Unfortunately, the composition determined without standards was never critically examined and became a part of accepted WHO and ATSDR Aroclor compositions. Consequently, Wolff et al. (1992) qualified their designation of CB 95 as a significant component in exposed workers. Although CB 95 is quite labile, it certainly could rival CB 18 which is a transient major component in workers directly exposed to former commercial mixtures (Luotamo et al., 1991; 1993; Maroni et al., 1981).

With all 209 congeners now available as standards, the carefully derived profiles of Frame et al. (1996a; 1996b), and the listing of retention times of all congeners on several columns (Bolger et al., 1995; Vetter and Luckas, 1995; Frame, 1997), errors in **Appendix Table 5** are also probable.

One of the most complete assignments of structures to low resolution and high resolution peaks can be found in Brown (1994). Many older low resolution studies contain valuable information that is more readily interpreted with more complete and accurate congener designations. Some of these designations are included in **Table 3.3** to facilitate interpretation of older literature.

It is hoped that this exercise will caution readers against casual and literal interpretations of residue reports.

3.2.1. Data Presentation

Attempting to gain insights from **Appendix Table 5** would have been much more difficult if the data presentation in **Table 3.1** (by peak and structure) had not been re-ordered by Ballschmiter and Zell (1987) (B & Z) numbers. The official IUPAC numbering system (**Appendix Table 1**) is cumbersome. Modifications such as presenting ring designations separately (e.g. Bush et al., 1985a; Hansen, 1987) and deleting commas (since only single digits are possible) (e.g. Brown, 1994; Frame et al., 1996; **Appendix Table 2**) make it easier to visualize the substitution pattern (**Figure 1.1**) and/or derive B & Z numbers. Merely arranging the peak designations and structures, without regard to analytical data, can be overwhelming. Editing oversights alone can further confuse the data and these errors may never be apparent unless one is looking for a specific CB.

We have long been intrigued by the 2,3,6-chlorinated congeners and the 13.6% tri-*ortho* CBs found in human tissues by Jensen and Sundstrom (1974) maintained this interest even though they seemed to disappear from many reported residue profiles. The focus on coplanar AhR agonists further contributed to indifference to their presence or absence.[1] Thus, partly from force of habit, residue reports are routinely

---

[1] Like the demonstration of selective binding of 2,3,6-CBs in lung (Brandt et al., 1981), the potency of CB 84 as an embryotoxicant in chickens as well as a depressor of microsomal oxidase activity (see Hansen, 1979; 1987) received little attention. In fact, in spite of input from USFDA and USDA scientists with first-hand knowledge, a DHEW Subcommittee (1978) incorrectly declared CB 84 to be nontoxic.

Table 3.3. More recent (than Table 3.1) comparisons of packed column peaks with high resolution gc (HRGC) designations. Additional details and clarifications in Brown, 1994.

Packed Column RRT[a]	B&Z No.	High Resolution Peak		
		Brown[b]	Chen[c]	Wolff[d]
0.28	15	14-2		1A
0.28	17,18	14-3,15-3		
0.37	31,28	23,24	4-3	1
0.47	47,49,52	31,32,33	6-4	2,3,4
0.54	44	37		5
0.58	64,71	39		
0.70a	74	46	9-4a	6
	70	47	9-4b	7
0.70b	66,95	48	9-4c	8,9
0.78	60	50		13
0.84	84	51		14
0.84	101,99	53,54	11-5-b,c	11,12
0.99	97	57	12-5-a	16
1.00	**DDE**	59		
1.01	87	58	12-5-b	17
1.05	110,77	61	13-5-a	18
1.25	118,149	69	15-5a	20
1.46	153,146,132	73,75	16-5	22
1.60	105,141,137	74,77,79	16-6	23
1.74	138,163	80,82	18-6	25
2.03	128,187,183	88,89,90	19-6	26,29
2.32	174,177	93,94		30,31,27
2.45	156	95	20-6	32
2.86	180	102	21-7	33
3.32	170,190	106,107	22-7	35

[a] Retention time relative to *p,p'*-DDE. Unit comparison (1.00) is used rather than percent (100) to avoid confusion with B&Z Numbers.
[b] Brown et al., 1989; Brown, 1994.
[c] Chen et al., 1982.
[d] Wolff et al., 1982.

scanned for CBs 84, 95, 110, and 149. This prelude is to help dispel the notion that the author is picking on an old friend. After having read several of Brian's articles (Bush et al., 1982; 1985a; 1985b; 1989) multiple times, we discussed congener assignments and he concurred that the changes in **Appendix Table 5** were probably justified.

More recent profiles by this lab have corrected the peak assignments (e.g., Kodavanti et al., 1999). Schwartz et al. (1987) also mis-assigned CB 82 to the CB 110/77 peak, but this group generally discounts non-TCDD-like congeners anyway (Stalling et al., 1979). In other cases CBs 82/110 may appear to be reversed in parallel sample sets (Baker and Eisenreich 1990; **Table 5.6**). Some of the confusion may have resulted from omission of CB 110 in early chromatographs (Ballschmiter and Zell, 1980; Zell and Ballschmiter, 1980); CB 110 was presented in later work, however (Ballschmiter et al., 1987). Other errors may be subsequently corrected, but this does not insure wide recognition. An experienced analyst would probably have recognized that "CB 121" was actually CB 74 (Luotamo et al., 1991); however, some of us may not have read the subsequent series of papers.

The reader must also note the objectives of the work. This author was in the process of emphatically questioning the PCB profiles presented by MacDonald et al. (1998). A more careful reading revealed that the extracts were fractionated and di- and tri-*ortho* CBs were discarded. In a previous review (Hansen, 1998), a methods-development paper (Duarte-Davidson et al., 1991) was incorrectly cited and the reported high proportions of CBs 126 and 188 were questioned. The definitive papers, however, did not include these congeners. CBs 15 and 119 were also deleted in the definitive studies (Duarte-Davidson et al., 1994a; 1994b).

Going back to Bush et al. (1985a) recently, I detected for the *first time* that CB 149 is listed both as 2,3,6,2',4',5' and 2,4,5,2',3',6' in their Figure 1 and incorrectly designated as IUPAC 131 in their Table 1 (by the way, the first CB 48 is really CB 49). A less justified editing oversight was the labelling of CB 66 as CB 44 directly above the correct designation (**Figure 1**, Li and Hansen, 1996a). If data organization can render a manuscript uneditable by the authors, it is not surprising that there is some confusion regarding the complex data generated by high resolution PCB analysis.

Thus, data presentation plays a very major role in interpretation. Peak sequences are important for qualitative evaluation of structure designations, but the sequence is unique to methods and conditions in each laboratory. It is frequently difficult to compare studies directly without considerable effort, and this leads to misinterpretations and misunderstandings. Some standardization of data presentation is appropriate now that high resolution analyses provide a potentially incomprehensible data set. Most of the data presented in this monograph have been rearranged by B & Z number. In addition, since qualitative interpretations are stressed, the congener amounts have generally been converted to percent of quantified peaks as suggested by Niimi and Oliver (1988).

The type of data presentation adopted for this monograph suits the objectives. Final standardization of data presentation will, of course, be up to the analysts and, perhaps, regulators. It is hoped that this need for standardized and simplified reporting will be taken seriously.

3.3. Selection of Analytes

"A problem with congener-specific analysis is the resulting information overload" (Erickson, 1997), but limiting analytes to a prescribed list was considered a "retrograde step since a balanced data set of a wider selection of CBs which have different biological activity would be intrinsically more useful with very little additional effort on the part of the analyst" (Hess et al., 1995). A considerable amount of valuable information is lost when PCB residue reports are abridged to include only "major" congeners or presented only as a total resembling a former commercial mixture.

The information lost was "in a large part a result of the rote mentality imposed by prescriptive standard methods such as the SW-846 Method 8080...." (Erickson, 1997). Low resolution studies not answerable to EPA generally presented all major peaks as a percentage of the Aroclor standard, preserving valuable information. Even the high resolution capillary method introduced by US EPA in 1990 still favored *quantitation* as Aroclor standards. A new generation of EPA scientists studying congener-specific biological actions are now echoing the 2-decade pleas to replace Aroclor equivalents with congener-specific residue data. Recently, specific-congener analyses were considered acceptable (USEPA, 1993; **Section 5.1.2**).

Most analysts prefer to let the results direct their choice of analytes. The experienced analyst is alert for overlooked and/or unanticipated analytes. For example, Tuinstra et al. (1994) added CB 99 to their list of analytes when it was found to be consistently present even though it was not an initial target. Bedard and May (1996) and Chiarenzelli et al. (1996) demonstrated that some rare substitution patterns can be introduced by anaerobic dechlorination (**Chapters 4 and 5**). Other rare CBs can be formed by processes such as incineration and chlorine bleaching of pulp (**Chapter 4**). Laboratory air can contain relatively high PCB levels which may be introduced into post-sampling profiles (Wallace et al., 1996). Storage, synthesis, or use of unique CBs in the laboratory may create some surprises. Note that recent regulations emphatically discourage PCB synthesis.

With the complex PCB residues encountered and associated interferences, some additional toxicological guidance is also appropriate. For example, the high potencies of coplanar CBs 77, 81, 126 and 169 made it imperative to develop methods to quantitate the very low levels in environmental samples. Difficult-to-resolve peaks may not warrant the doubling or tripling of effort required for resolution if there are no compelling biological reasons to know exact proportions. Each matrix has unique interferents which may increase detection limits for different analytes; if the trace levels so masked are of little relative importance, the methods can be streamlined and the reports simplified.

Guidance was suggested by Jones (1988) with a reasonably broad profile of congeners accounting for about 70% of most environmental burdens and including the low-occurring coplanar congeners. A year later, a more restrictive approach was suggested (McFarland and Clarke, 1989). Unfortunately, their "environmental occurrence" was based mostly on 2 articles. The capillary column analysis of a single human milk sample was the basis for their human tissue occurrence which was used to negate the importance of congeners previously found in human fat and in the environmental samples. The timing was right, however, and focus on non- and mono-*ortho* CBs was readily accepted and carried to further extremes by subsequent studies which felt obliged only to report 3 dominant congeners, coplanar PCBs and, sometimes, mono-*ortho*-CBs.

Wolff et al. (1997) suggested modification of the entrenched self-limiting "lists" and this was expanded by Hansen (1998). If only traditional CBs are reported, the misconception will be that these are the only ones of importance. With this misconception, there would seem to be little reason to study the biological actions of other congeners. With no reference to biological activity, there is then little need to resolve or report these "non-traditional" congeners. If this self-perpetuating abridgment continues, it will be impossible to accurately associate biologically plausible medical outcomes with PCB exposure (Wolff et al., 1997; Hansen, 1998). It will likewise be impossible to exonerate PCBs from implied associations with adverse medical outcomes without more complete residue and toxicity data.

The toxicological data base is expanding and the analytical data for possible associations are readily obtainable. Now these data bases must be merged to determine whether or not PCBs present a serious threat to wildlife and public health. For surveys of contamination and trends, large numbers of complex samples must be analyzed and estimations based on prominent marker peaks may suffice; however, it is critical to recognize the objectives so that ***not reported*** is never equated with ***not present***. The remainder of the monograph suggests congeners which may have been overlooked in abridged reports, but are implicated in biological actions which ***may*** lead to adverse biological responses.

4 ATMOSPHERIC TRANSPORT

After several decades of manufacture and use in open systems, PCBs are globally distributed. Considerable variation can be found in the estimates of total global production, with amounts varying from 6.1 to 12.0 x 10^{11} g (1.3-2.5 billion pounds) (Erickson, 1997). The lower figure includes limited production estimates into the 1970s, while the higher figure extends to 1980 and includes former USSR production of Soval on into the 1990s. Erickson (1997) offers considerable discussion and insight into the derivation and probable accuracy of these values.

This chapter introduces global disposition of these PCBs, mainly by atmospheric transport. Some of the properties which favor volatility (**Chapter 2**) are used to emphasize the congener-specificity of airborne PCBs. Sources and profiles are compared and transfer to biota is examined. More detail regarding PCB disposition in global and local environments is provided in **Chapter 5**.

4.1. Global Compartments

There are geographically-related "ambient" levels of contamination in air, waters, and biota (See ATSDR, 1997; Erickson, 1997). There are also certain "hot spots" containing elevated levels. Repeated cycling of PCBs into and out of accessible environmental reservoirs occurs so that PCBs are dispersing from highly contaminated to less contaminated areas; thus, the differences among "contaminated" and "ambient" sites are diminishing (Eisler, 1986; Erickson, 1997). The most important compartment in this dispersion is the atmosphere (e.g., Bidleman and Olney, 1974; Eisenreich et al, 1981; 1992; Jones et al., 1992; 1995).

It must be understood that only a small portion of the total PCBs available are airborne at any one time. There are ranges of estimates, but total fractions in the atmosphere are generally less than 0.05% of total global PCB (**Chapter 5**).

Congener-specific relative distributions among major environmental compartments calculated from fugacity based models (Mackay et al., 1992) are presented in **Table 4.1**. The "average" global atmospheric load of about 0.03% is lower in remote areas and generally higher over the open ocean (NAS, 1979; Tatsugawa and Tanabe, 1984). Further discussion of these distributions is presented in **Chapter 5**.

Table 4.1. Congener-specific distributions as predicted from Level I and Level III fugacity calculations (Mackay et al., 1992). Calculations are based on a 100,000 km^2 landmass with a 10,000 km^2 water surface with an average depth of 20 m. Accessible soil has a volume of 10^9 m^3 and bottom sediment a volume of 10^8 m^3. Level I includes 100,000 kg of PCB at equilibrium. Level III calculations are based on 600, 300, and 100 kg PCB/h discharged into air, water and soil, respectively, and incorporate more intermedia transport rates.

CB	Distribution (%)				Loss (kg/h)	
	Air	Water	Soil	Sediment	React	Advect
Level I (Input 100,000 kg)						
18	4.88	0.26	92.7	2.06		
52	0.84	0.09	96.8	2.15		
101	0.31	0.04	97.4	2.16		
128	0.03	0.01	97.7	2.17		
Level III (Input 1000 kg/h)						
18	0.7	1.03	79.5	18.8	204	672
52	0.5	0.77	68.1	30.6	141	718
101	0.3	0.55	68.9	30.3	239	667
128	0.1	0.28	78.4	21.2	428	479

4.2 Atmospheric PCB Levels

In the 1970s, oceanic and rural continental atmospheric PCBs ranged from 0.002-1.6 ng/m^3 while urban atmospheres contained 0.5-36 ng/m^3 (NAS, 1979). Measured atmospheric levels are highly seasonal and temperature dependent (Hermanson and Hites, 1989; Lewis et al., 1985; Monosmith and Hermanson, 1996; Tanabe et al., 1983; Vermette et al., 1995; Wallace and Hites, 1996) so it is difficult to compare older studies directly. Regional differences, however, were quite marked. NAS selected a value of 5.0 ng/m^3 as representative of average atmospheric concentrations over urban areas in the US (10% of area) and this seems reasonable. The value of 0.05 ng/m^3 selected to represent the 90% of the atmosphere over rural, oceanic and remote areas seemed too low. More recent values suggest a smaller difference between average rural and average urban levels (see also **Table 5.4**). This may reflect more critical sampling and analysis and/or more thorough mixing as point sources are decreased.

Total atmospheric PCBs are rapidly dispersed, but the minimum 10-fold difference between average urban and rural air demonstrates the maintenance of localized contaminant levels. Different air masses at heights up to 10,000 ft (3.05 km) have been shown to differ by factors of 2- to 5-fold (Knap and Binkley, 1991). General levels were 0.2-0.4 ng/m^3, but the air sampled at 9-10,000 ft over the NY Adirondacks in 1985 contained as much as 1.4 ng/m^3 total PCB. Airborne total PCBs over the Atlantic Ocean may differ by an order of magnitude (Schreitmüller and Ballschmiter, 1993; **Chapter 5**).

At rooftop height, local atmospheric PCBs still reflect local ground level PCB contamination during long-term monitoring (Hermanson and Hites, 1989). Clean-up efforts of localized sources reduce ground-level air concentrations, but do not necessarily attenuate local atmospheric contamination at breathing height (180 cm) (Lewis et al., 1985) or rooftop height (Panshin and Hites, 1994) even after several years. Calculated inputs from point or localized sources cannot account for the maintenance of high atmospheric levels (Murphy et al., 1985). Although residence times of individual molecules can be reasonably accurate, the dynamic nature of the atmospheric reservoir defies cleansing. Once airborne, point sources may soon become non-point source reservoirs as PCBs disperse, deposit and re-volatilize.

Re-appearance of some fractions is occurring in the Northern latitudes due to temperature-dependent long range partitioning predicted in the early 1970s (Rappe, 1974; Goldberg, 1975; Ottar, 1981). The process has recently been refined and referred to as "global fractionation" (Wania and Mackay, 1993; Muir et al., 1996; **Section 4.4**). The lower chlorinated and *ortho*-enriched PCB congeners have long been recognized as the most mobile due to greater vaporization rates and lower binding affinities for particulates (**Chapter 2**). This is readily confirmed by theoretical considerations (Falconer and Bidleman, 1994) as well as the many observations cited in this chapter.

4.3. Dispersion of airborne PCBs

Even though particle-bound and vapor phase PCBs in the atmosphere at any one time represent a low proportion of the total "accessible" global burden, the atmosphere is the major route of global transport. The foundation for current concepts of atmospheric transport and differential partitioning of organochlorines might be traced to Bidleman and Olney (1974). It was recognized that the majority of PCBs and DDT were actually in the vapor phase so that previous estimates of residence times based on particle-bound residues were up to 20 times too long. Concentrations were higher in the ocean surface microlayer and decreased with depths. The atmospheric residence time was estimated at 40 to 50 days for total PCB, surprisingly similar to recent calculations of 49 $\pm$ 16 days for some 80 congeners detected in air samples (Panshin and Hites, 1994).

4.3.1 Differential Dispersion

The enrichment in lower chlorinated congeners of atmospheric PCB mixtures (Bidleman and Olney, 1974) has been repeatedly confirmed and trichlorobiphenyls are the most prevalent homolog group (Atlas and Giam, 1981; Tanabe et al., 1983). More recently, specific congener analysis has permitted further refinements in the characterization of airborne PCBs. The vapor phase is also enriched (relative to commercial mixtures) in di- and tri-*ortho* congeners within each homolog group due to greater vapor pressures and a lesser tendency to bind to particulates (Falconer and Bidleman, 1994; Hippelein and McLachlan, 1998; Jones et al., 1992; Monosmith and Hermanson, 1996; Muir et al., 1996; Panshin and Hites, 1994; Wania and Mackay, 1993; Simcik et al., 1998). Properties associated with vaporization of CBs have been presented in **Appendix Table 3** and compared with some co-pollutants in **Table 4.2**. Trichlorobenzenes were commonly used in transformer askarels with Aroclors 1254 and 1260 (Erickson, 1997), so sample chlorobenzene and PCDFs are included for perspectives. Some comparative specific congener reports of vapor phase PCBs are presented in **Appendix Table 6**..

Being much less volatile, coplanar non-*ortho* and higher chlorinated PCBs and, especially, PCDDs and PCDFs are present at very low proportions in the vapor phase and tend to be removed from the atmosphere with particulates by both wet and dry deposition. ATSDR (1994) suggests half-lives of PCDFs in the air as about 10 hours with most of the loss due to deposition. Vapor phase PCBs, PCDDs and PCDFs may also be lost through reactions with hydroxyl radicals with half-lives as short as 1-2 days (Anderson and Hites, 1996; Pennise and Kamens, 1996). An organic hydrogen donor appears to be essential, so that little or no reactivity is observed for these compounds when residing on particles generated by high-temperature combustion (Pennise and Kamens, 1996). The 2-phase removal of PCDDs/Fs from the atmosphere is congener-specific: lower chlorinated congeners are removed mainly from the vapor phase by reaction with OH radicals while dry deposition is more important for higher chlorinated congeners. The OH radical reactions are less important at lower temperatures (Brubaker and Hites 1997). Higher chlorinated CBs would behave similarly, although not quite so rapidly. Some common PCBs such as CBs 28, 44, 47, 110 (half-lives 11-19 days) and, especially, CB 95 (half-life 34 days) are much less reactive even in the vapor phase than are PCDDs and PCDFs (Anderson and Hites, 1996).

Once particle-bound, these compounds are resistant to release and enter water bodies by deposition as well as move with wind- and water-eroded soils until they are fixed in sediments or deeper soils. Water as a recipient and sediment as an ultimate repository for the less volatile CAs is confirmed by the ability to predict and model historic trends using data from date-able sediment cores. They do not revolatilize as readily as do some of the lower-chlorinated and *ortho*-enriched PCBs. Thus, depending on meteorological conditions, the PCDDs, PCDFs, and some PCBs would

Table 4.2. Relative water solubilities and vapor pressure estimates for selected chlorinated aromatics (ATSDR, 1994; Mackay et al., 1992[a]; **Table 2.3**).

Congener	Substitution	Water Sol. (µg/L)	Vapor Press (Pa @ 25°C)
DiCl-Benzene	1,4	83000	90
TriCl-Benzene	1,2,4	40000	48
TetraCl-Benzene	1,2,4,5	7800	9.6
CB 2	3	2500	1.00
CB 4	2,2'	1000	0.26
CB 5	2,3	1000	0.15
CB 8	2,4'	1000	0.15
CB 10	2,6	1400	0.33
CB 15	4,4'	60	0.0048
CB 18	2,2',5	400	0.14
CB 26	2,3',5	250	0.035
CB 28	2,4,4'	160	0.026
CB 47	2,2',4,4'	90	0.0054
CB 52	2,2',5,5'	30	0.0049
CB 66	2,3',4,4'	40	0.001
CB 77	3,3',4,4'	1	0.000059
CB 95	2,2',3,5',6	34	0.006
CB 99	2,2',4,4',5	11	0.003
CB 101	2,2',4,5,5'	10	0.002
CB 110	2,3,3',4',6	10	0.003
CB 138	2,2',3,4,4',5'	1.6	0.0005
CB 153	2,2',4,4',5,5'	1	0.00012
4-CDF[b]	2,3,7,8	0.42	0.0000009
5-CDF[b]	2,3,4,7,8	0.24	0.0000002
6-CDF[b]	1,2,3,4,7,8	0.008	0.00000006
7-CDF[b]	1,2,3,4,6,7,8	0.014	0.00000002

[a] Some values were averaged from dissimilar presented values if a best value was not selected by the authors.

deposit onto soils, plants and bodies of water closer to any point of origin. The more volatile congeners would tend to move farther, even if deposited and re-volatilized, helping to explain enrichment of these congeners in snow pack and biota at colder latitudes.

Although average collective vapor pressures vary dramatically among PCB homolog groups (DiCB = 0.2 > TetraCB = 0.004 > HexaCB = 0.0001) (Wania and Mackay, 1993), isomers within a homolog group may easily vary by an order of magnitude so that some heavier *ortho*-chlorinated congeners may be more volatile than lighter *para*-chlorinated coplanar congeners (compare CBs 66, 77, 95 and 110 in **Table 2.2, Table 4.2** and **Appendix Table 3**). CBs 66/95 and 77/110 co-elute on many GC columns and can usually be distinguished only by masses. The congener composition of vapor phase PCBs, of course, is dependent on the congeners present in the source as well as the relative vapor pressures of the congeners present.

Some congeners which are not important constituents of former commercial mixtures may occur due to *in situ* processes such as incineration (Ballschmiter et al., 1987), chlorine bleaching of wood pulp (MacDonald et al., 1998), or anaerobic dechlorination (discussed in **Chapter 5**). CBs 24 and 27, which are minor components of Aroclors (Frame et al., 1996; **Appendix Table 2**), are found in significant proportions in soil, air and dust associated with a PCB landfill (Hansen et al., 1997) and in the congener mix volatilized from contaminated sediment (Chiarenzelli et al., 1996; 1997). They are not found at appreciable levels in the volatilization mixture from subaqueous sand to which Aroclors have been added (Chiarenzelli et al., 1997), leading the authors to conclude that they were probably generated in the sediment from anaerobic dechlorination. Note that the authors attribute 9% of the PCB volatilized from Aroclor 1260 to CB 66 (Chiarenzelli et al., 1997); however, this congener is a minor component of Aroclor 1260 (0.03%) compared to the co-eluting CB 95 (2.5%) (Frame et al., 1996). CB 95 was not included as a reportable analyte, but is more volatile than CB 66 (**Table 4.2**).

An interesting observation from the previous papers was that volatilization of lower chlorinated and *ortho*-enriched congeners was greatly enhanced from wet vs dry sand or sediment. This was attributed to desorption followed by evaporation-enhanced volatilization as with steam distillation processes (Chiarenzelli et al., 1996; 1997). It is most likely that desorption processes played the major role, since it has long been known that chlorinated aromatics cannot be efficiently extracted from dehydrated matrices (Bertazzi et al., 1967; Armour and Burke, 1971) or eluted from substrates such as alumina without some deactivation of sorption sites with water (Erickson, 1997). The reminder of the importance of hydration in volatilization (and other losses) serves to reinforce the importance of vapor phase PCBs in global transport and caution against naive handling of samples and remediation wastes.

Another interesting recent observation is the binding of certain CB congeners to nonfilterable (submicron) particles in rainwater (Poster and Baker, 1996). Individual CBs appear to interact differently (independent of partition coefficients) with submicron particles than with macroparticulates. CB 153 was found equally distributed between filter-retained and nonfilter-retained particles during one rain event. CBs 16/32, 33/53, 70/76, 118 and 146 were decidedly enriched on nonfilterable particles while CBs 28/31, 174, and 187 were more enriched on filter-retained particles in the rain. This enhanced precipitation scavenging of certain CBs independent of commonly considered constants may also disrupt "ambient profiles" of atmospheric PCBs.

4.3.2 Differential Biological Consequences

The different types of CBs can cause broad spectra of sometimes overlapping biological effects (**Table 2.7**). The localized profiles of CB burdens would also be altered from anticipated "global equilibrium" profiles in recently impacted areas. Because of the greater potency of PCDDs, PCDFs and coplanar CBs for a defined constellation of effects and the much broader profile of biological responses to other PCBs, the characteristics of the emissions and the resulting plume(s) must be described before attempting to determine environmental impacts.

Vapor phase halogenated aromatics are important, not only because they are more mobile, but because they are more readily absorbed by vegetation for entry into the food chain (Lorber et al., 1994). Selective accumulation of lower chlorinated and *ortho*-enriched congeners in nonvascular plants, pasture herbage and produce has been recently confirmed in a variety of studies (e.g., Cullen et al., 1996; Dushenko et al., 1996; Hansen, 1998; Himberg and Pakarinen, 1994; Jones and Duarte-Davidson, 1997; Jones et al., 1992; 1995; Muir et al., 1993) (**Section 4.6**). Sample herbage-accumulated congener profiles are in **Appendix Table 7**.

The more potent coplanar (mono- and non-*ortho*) congeners are more readily particle-bound and less available in the vapor phase, rendering them less mobile and less bioavailable. Of course, none of the chlorinated aromatic (and other semivolatile) air pollutants can be considered in a vacuum since toxic interactions are likely; however, having included PCDDs and PCDFs in the general discussion is adequate to stress the points most relevant to the topic of this book.

4.4. Global Fractionation

Ambient atmospheric profiles of PCBs are most dependent on volatilities, degradation and binding to larger particulates and the less chlorinated and *ortho*-rich CBs dominate (e.g., **Appendix Table 6**). Once airborne, the net effect of grossly different volatilities is similar to fractional distillation of a complex mixture. In the first meter above contaminated soils, relative contributions of the homolog groups change (**Table 4.3**). Tri- and tetraCBs disperse most rapidly so that by 60 cm above contaminated soils they are at concentrations 20-24% as high as at 15 cm and only 11-14% of 15 cm values at 90 cm above the soil. Penta- and hexaCBs still contribute 32-36% of 15 cm values at 60 cm and 20-24% of 15 cm values at 90 cm. In other words, triCBs and *ortho*-rich congeners dominate the vapor PCB profile near the soil surface; however, they are more rapidly dispersed and contribute relatively less to the total PCB profile at modestly increasing heights directly above the soil. At greater heights and distances from the source, the less mobile congeners are more depleted and the mobile CBs again dominate the profile.

The lower chlorinated and *ortho*-enriched CBs would not only move more rapidly and remain in the vapor phase longer, but would revolatilize from soils and waters more readily for continued migration. At temperature extremes, the relative differences in mobility between CBs of modest vs. low volatility would be even greater. The distinctions would be more apparent in remote regions where releases of unweathered PCBs would have a smaller impact. Between latitudes 50-70° N, tri-CBs start to increase in biota such as burbot liver with latitude, while tetra- and penta-CBs are nearly constant; on the other hand, proportions of hexa- and hepta-CBs decrease dramatically with increasing latitude in these fish sampled from remote regions (Muir et al., 1990; Wania and Mackay, 1993). These relationships are opposite to what would be expected based on persistence alone. Likewise in Antarctica, di-, tri- and tetra-chlorinated congeners decidedly dominate the PCB profile in the hydrosphere, but especially in the atmosphere (Tanabe et al., 1983).

Further support for the fractionation theory can be gleaned from studies of PCB accumulation by mosses in remote regions of Europe and North America (Himberg et al., 1994; Lead et al., 1996; Muir et al., 1993). Recent studies continue to confirm the increased dominance of lower chlorinated and *ortho*-enriched PCBs and lower chlorinated PCDD/Fs in both vapor phase and biota of colder latitudes (Halsall et al., 1997; Jones and Duarte-Davidson, 1997; Thomas et al., 1997).

It should be clarified that ***most*** of the global diCBs, triCBs and tetraCBs will ***not*** eventually deposit in the polar regions (Wania and Mackay, 1996). These homologs readily volatilize in temperate regions, partitioning out of the air at colder temperatures; however, they are also readily retained and degraded en route. Polar terrestrial environments have limited retention capacities so that much of the mobile PCB load transfers to aquatic and marine environments. Nevertheless, net transfer of only small proportions of PCBs from tropical and temperate areas to the relatively much smaller and less biota-buffered polar regions could be highly significant. The updated models, specific properties and categories, and misinterpretations of "global fractionation" are succinctly and thoroughly addressed by Wania and Mackay (1996).

Table 4.3. Comparison of the PCB homolog composition of 18 pooled 24-h high volume vapor phase extracts (Hansen et al., 1997) with the single 1 h air sample from the same height used to determine the vertical gradient. The vertical gradient was determined by comparing the congener concentrations at 60 and 90 cm with those of the reference level (15 cm) (Vermette et al., 1995; **Appendix Table 8**).

Homolog Group and Selected CBs	% Total PCB at 15 cm		% of value at 15 cm	
	Pooled	Single	60 cm	90 cm
Trichloro				
16+32	7.43	10.28	16.28	9.80
18	7.39	8.35	36.95	24.43
28+31	19.93	20.57	20.00	12.00
AVG (n = 5)	*8.4*	*9.4*	*23.0 ± 7.4*	*14.1 ± 5.3*
AVG (n = 4)[a]	8.63	9.64	**19.5 ± 2.6**	**11.5 ± 1.4**
Tetrachloro				
44	4.67	5.27	20.86	10.86
52	6.43	7.57	19.89	13.70
66	2.86	2.22	29.07	16.04
AVG (n = 12)	2.87	3.13	**23.8 ± 5.0**	**13.6 ± 3.1**
Pentachloro				
95	1.84	1.50	32.14	17.72
99	0.87	0.90	31.93	18.72
101	2.58	3.17	35.04	22.06
AVG (n = 12)	0.92	0.91	**32.5 ± 1.3**	**20.2 ± 2.6**
Hexachloro				
138+163	0.42	0.37	34.88	23.92
149	0.58	0.59	33.82	20.90
153	0.41	0.34	38.47	24.03
AVG (n = 9)	0.23	0.23	**35.5 ± 4.0**	**22.6 ± 2.2**
TOTAL PCB			24.0%	14.3%

[a]Excluding CB 18 which is an obvious outlier for % 15 cm comparisons.

On a smaller scale, accumulation of lower chlorinated PCBs has been detected in remote Rocky Mountain lakes. These CBs are presumed to be derived from higher chlorinated mixtures because of their greater atmospheric mobilities (Heit et al., 1984). Desert bighorn sheep from remote areas in California had low pesticide and moderate PCB levels; however, the single sample from Los Angeles County had disproportionately (7-fold) higher PCB than pesticide residues (Turner, 1978), presumably due to consistently higher PCB levels in urban atmospheres (e.g., NAS, 1979; Panshin and Hites, 1994; Simcik et al., 1998; **Section 4.2**).

Even more localized are diurnal variations in vapor phase PCBs; the highest concentrations are recorded during periods of the highest temperatures (Wallace and Hites, 1996). Specific congeners were measured, but results were not presented. On a global scale, the more volatile PCBs are being "distilled" into the colder regions while PCDD/Fs and higher chlorinated/*ortho*-poor congeners tend to deposit with particulates and ultimately be buried in sediments. On a more local and short-term scale, soils, plants, and waters closer to emission sources would contain relatively higher levels of TEQ-contributing CAs while the periphery of affected areas would contain greater proportions of the more volatile congeners with totally different biological actions.

4.5. Character of Localized Emissions

Uncontrolled landfills have long been known to be significant sources of airborne PCBs (Hermanson and Hites, 1989; Lewis et al., 1985; Murphy et al., 1985; NAS, 1979). The behavior of PCBs above and near releases from uncontrolled landfills can be predictive of point source releases and help to understand atmospheric profiles.

Vent gases from 5 different landfills contained 37-390 ng PCB/m^3 total air; when normalized to methane emissions to compensate for extraneous air, PCBs were vented at 70-500 (mean 285) ng PCB/m^3 of methane (Murphy et al., 1985). The main vent in a controlled PCB landfill released about 120,000 ng/m^3 (Lewis et al., 1985). Ambient air levels near this landfill remained below 6 ng/m^3 since this only amounted to a release of about 0.04 mg/h. The air concentrations over more diffuse "hot spots" (visible leaking capacitors) in 3 uncontrolled landfills were about 400-40,000 ng/m^3 and as high as 193,000 ng/m^3 (Lewis et al., 1985). Dissipation was rapid (**Table 4.4**), but the packed column profile of PCBs resembled that of Aroclor 1242 and did not appear to change appreciably between 2 and 120 cm above the landfill surface. More recent studies using specific PCB congener analysis demonstrate that the more volatile congeners actually do dissipate somewhat more rapidly (See **Table 4.3** and below).

Table 4.4. Total PCB concentrations determined in high volume samplers at various heights above contaminated and surface-cleaned electrical landfills (Lewis et al., 1985).

Sample Height	Site 1a			Site 1b	
	Oct	Jul	Apr (Cleaned)	Jul	Apr (Cleaned)
2 cm[a]	396 ng/m^3	815 ng/m^3	2.75 ng/m^3	855 ng/m^3	16.4 ng/m^3
30 cm[b]	7.6%	10.8%	53%	16%	60%
60 cm[b]	3.4	5.4	42	6.0	15
120 cm[b]	2.2	2.9	38	2.1	15
180 cm[b]	0.5	0.5	18	1.1	12

[a]Mid-range or average concentration (ng/m^3) at lowest height sampled.
[b]Percent of maximum concentration at this height.

The dispersion of low concentrations appears to be much less marked (**Table 4.4**); however, the contribution of ambient levels to measured concentrations becomes more important at these low levels. More recent studies employing specific congener analyses indicate that the profile of congeners changes modestly within the first meter of dispersion (Hansen et al., 1997; Vermette et al., 1995; **Table 4.3**; **Appendix Table 8**). The mono- and dichloro-CBs as well as CB 18 appear to reside in the 90 cm "plume" longer than other tri- and tetrachloro-CBs (**Appendix Table 8**); however, this may be due to a greater rate of input. The *ortho*-poor/di-*para* tetra-CBs 66, 77 and 81 are decidedly more slowly dispersed than the other tetra-CBs (**Appendix Table 8**).

A small 2.5 acre (1 hectare) electrical dump in Southern Illinois had been studied extensively prior to recent remediation. Airborne PCBs have been compared to other profiles in **Appendix Table 6**. The relative dispersion of homologs has been presented in **Table 4.3**. The soil and overlying vapor phase PCBs from this weathered electrical landfill resembled, but were not identical to, Aroclor 1248 (**Table 4.5**; Li and Hansen, 1996; Hansen et al., 1997). The PCB congeners extracted from superficial dust and debris were somewhat more highly chlorinated than those from the air and subsurface soil. This could be attributed to the enhanced volatility of PCBs from wet substrates (Chiarenzelli et al., 1996; 1997) since the "dust and debris" would contain a large amount of organic matter and experience greater variations in hydration-dehydration than would the underlying soil. The hydration-enhanced volatilization provides the "trapping" mechanism the authors proposed to explain the higher chlorine content of the dust PCBs (Hansen et al., 1997).

A localized release of over 400 mg/h combined PCDDs + PCDFs and over 500 g/h each of total PCBs and total trichlorobenzenes was recently experienced in a sparsely populated region in Canada. The reported PCB homolog composition of the process gas/flue gas stack emission sample extract (Wellington Laboratories, 1997) is compared to other PCB mixtures in **Table 4.5**. The stack sample closely resembled a mixture of Aroclors 1254 and 1260 in homolog distribution. This short-term stack release is compared to a more continuous release of PCBs and PCDD/Fs from the retired electrical equipment landfill in Southern Illinois (Sangamo Landfill).

Table 4.5. PCB homolog percent composition of an accidental transformer furnace stack release compared to Aroclor 1254 and PCBs from the Sangamo Electric Landfill (SLF)[a]. T = Trace. Major homolog groups and TCDD equivalents (TEQs) are in bold.

Homolog	Stack Release	Aroclor 1254	SLF Air	SLF Dust	SLF Soil
Mono-Cl	0.04				
Di-Cl	0.18	< 0.1	1.2	0.2	0.9
Tri-Cl	1.36	2	**45.5**	19.4	**38.0**
Tetra-Cl	12.46	17	**39.1**	**39.0**	**41.3**
Penta-Cl	**29.08**	**49**	11.9	**25.8**	12.1
Hexa-Cl	**33.83**	**28**	2.2	12.8	5.4
Hepta-Cl	18.40	4	0.06	2.5	2.0
Octa-Cl	4.45	T		0.25	0.23
Nona-Cl	0.33	T		0.02	0.02
Deca-Cl	0.004			0.004	0.002
TEQ/PCB (μg/mg)	**0.54**	**0.14**	**0.11**	**0.51**	**0.62**

[a]Original homolog compositions of SLF samples (Li and Hansen, 1996) have been corrected according to Hansen et al., 1997.

The equivalencies to 2,3,7,8-TCDD (TEQs) compared to total PCB concentrations were similar for the stack emissions and the landfill soil and dust (**Table 4.5**). Landfill residues were enriched in PCDFs due to open burning during the 1950s and 1960s (Hansen and O'Keefe, 1996) while the stack emissions were similarly enriched due to the heating process. The vapor phase PCBs over the landfill contained lower TEQ concentrations more reflective of Aroclor 1254 because of the lower volatility of PCDDs, PCDFs (ATSDR, 1994) and coplanar PCBs (Wania and Mackay, 1993) (**Table 2.3; Table 4.2**).

Because of the different volatilities of the various chlorinated aromatics emitted from the stack, the composition of the mixture will change as it moves downwind. The more highly chlorinated PCBs and, especially, the PCDDs and PCDFs are much less volatile than more lightly chlorinated and multi-*ortho* chlorinated PCBs (ATSDR 1994; 1997; Mackay et al., 1992; Wania and Mackay, 1993; **Section 4.3**). This is a much smaller scale than the global fractionation concept and even than the more localized movement of PCBs to remote Rocky Mountain lakes (**Section 4.4**); however, it is a larger scale than the definitive diurnal variations in atmospheric PCBs reported by Wallace and Hites (1996) and should cause a transient localized impact. Preliminary studies in addition to required monitoring have helped to characterize the spatial and temporal trends (e.g., Blais et al., 1998; Chen et al., 1998; Froese et al., 1998; Gabos et al., 1998). It is hoped that additional investigations of this recent release will be possible.

4.6. Accumulation of Airborne PCBs by Vegetation

Early versions of the air → food models underpredicted because they failed to consider or underestimated the observation that herbage concentrations of, e.g. TCDD, "are overwhelmingly dominated by vapor transfers" (Lorber et al., 1994). This vapor-to-plant transfer has also led to a variety of studies using various plants as monitors for trends in CAs and PAHs (Gaggi et al., 1985; Focardi et al., 1991; Jones et al., 1992; Muir et al., 1993; Himberg and Pakarinen, 1994; Duschenko et al., 1996; Thomas et al., 1997). Some reported specific PCB congener contents of vegetation samples are presented in **Appendix Table 7.**

The greatest proportion of vapor transfer work has been done with PCDDs and PAHs, but similar principles apply to PCBs. There are obviously plant species differences in uptake and loss (McCrady et al., 1994). These are due to physical properties (Paterson et al., 1991), plant physiology (Sabljic et al., 1990), growth habits (Lorber et al., 1994), and cultivation practices (Jones and Duarte-Davidson, 1997).

Grasses are efficient scavengers of atmospheric CAs and actual transfer values can be determined from long-term simultaneous measurements of airborne and harvested grass concentrations of CAs (Jones and Duarte-Davidson, 1997; Thomas et al., 1997). The compound specific accumulation potential can be seen in **Table 4.6** in the volume of air required for the CB accumulated by each gram of grass harvested in 2-week intervals for 9 weeks. The higher accumulation of the more volatile PCBs results in a smaller volume of air represented in each g (dry weight, DW) of harvested grass. TCDD and TCDF accumulations are similar to air/grass ratios of the heptachlorobiphenyls, but are also compound-specific (Jones and Duarte-Davidson, 1997). Greater accumulation efficiencies are realized by optimized growth (Jones and Duarte-Davidson, 1997).

Table 4.6. Efficiency of grass scavenging of CAs from the atmosphere by comparing long-term air and grass concentrations (Thomas et al., 1997). The air/grass represents the amount of air required to achieve the CA concentration in grass.

Compound	Grass Conc (pg/g DW)	Air/Grass (m^3/g DW)
CB 18	115	8.5
CB 28	120	8.9
CB 52	30	7.1
CB66(95?)	30	10
CB 101	35	15
CB 118	30	26
CB 153	40	22
CB 180	35	55
CB 170	15	69
Total PCB	**900-2300**	
2378-TCDD	0.7	80
Total PCDD	**260**	
2378-TCDF	14	42
Total PCDF	**156**	

The lower chlorinated and *ortho*-enriched congeners, already at higher proportions in the vapor phase, are also more efficiently scavenged by other vegetation. Ombrotrophic mosses and lichens accumulate decidedly higher proportions (**Appendix Table** 7) because they purposefully scavenge the air for water and nutrients; vapor phase hydrophobic contaminants are accumulated coincidentally.

Protected grains have lower concentrations than do the leafy parts of the plants because internal transfer is minimal. Below-ground produce such as potatoes accumulate lower levels than do, e.g., tomatoes; tomatoes accumulate lower levels than leafy produce such as lettuce (Cullen et al., 1996). Whole carrots appear to accumulate generally higher relative levels than most other underground plants. Potatoes accumulate lower proportions of the more lightly chlorinated congeners than do tomatoes, but the di*ortho* pentaCBs 99, 101 and 110 are decidedly dominant in potato profiles (Cullen et al., 1996).

4.7. Other Environmental Pathways for Airborne PCBs

Of course, plant uptake is not the only source of vapor phase PCBs for biota. It has long been known that inhaled PCBs are efficiently absorbed (Treon et al., 1956; Benthe et al., 1972); however, food-chain contributions to the accumulation of airborne CAs is far more important due to biomagnification by plants and animals (Hattemer-Frey and Travis, 1991; Lorber et al., 1994). For TCDD, the relative contributions from fruits and vegetables (3.9%) estimated by Hattemer-Frey and Travis (1991) may be too low compared to inhalation (6.8%) (Lorber et al., 1994). Certainly, a major proportion of the animal product contribution to TCDD intake is indirectly through vapor and deposition contamination of forages (Hattemer-Frey and Travis, 1991; Lorber et al., 1994), so that the inhalation route becomes even less important, even for TCDD. For the more volatile and herbage-accumulated PCBs, the contributions would shift to an even greater proportion.

Several other examples of fauna-related PCB transport and transformation are available, but insects provide a suitable example. Insects have a tremendous potential for contaminant dispersion and mixture alteration by virtue of their large biomass, occupation of nearly every available niche, positions in most food webs and frequent occupation of different media in different life stages. The above observations and the fact that insect control can be achieved by both residual and aerosol pesticides testifies to their ability to absorb lipophilic chemicals from several sources in addition to the food chain.

Insects can accumulate significant amounts of vapor phase CAs directly and from contact with soil/air interchange surfaces as well as from consuming contaminated vegetation. Beetles (*Dyscinetus picipes*) collected from the Sangamo Landfill site (**Table 4.3, 4.5; Appendix Tables 6, 8**) contained 4-6 ppm PCB while beetles from a control site had no detectable (apparently less than 0.09 ppm) residues (McKee, 1992). House crickets (*Acheta domestica*) suspended (in a protected manner) about 1 cm above the soil at the same location accumulated 1-1.5 ppm PCB from aerosol exposure (Paine et al., 1993). Laboratory exposures of the crickets to Aroclors leading to residues less than 150 ppm did not cause acute toxicities. This is similar to results with house flies (*Musca domestica*) which tolerated relatively large topical doses of PCB congeners before experiencing decreases in longevity (Tehseen et al., 1992). Thus, insects can accumulate high residues from various sources with little adverse impact and readily transport, alter or contribute PCBs to the food chain.

Cuticular penetration is fairly rapid for *ortho*-rich congeners and slower for coplanar congeners (Li and McKee, 1992; Saghir et al., 1993; 1994). In fact, cuticular penetration in insects is by simultaneous rapid and slow absorption processes (Saghir et al., 1993; 1994; **Chapter 7**) which may be similar to absorption by plants with analogous trapping cuticles and rapid entry stomata/trachae.

Insects have active P450s which may be enhanced by inherent defenses against phytotoxins (Berenbaum et al., 1996; Ma et al., 1994) or acquired resistance to pesticides (Hodgson, 1985; Mullen and Scott, 1992). Significant biotransformation of PCBs occurs (Saghir et al., 1994; Saghir and Hansen, 1992) and this metabolism can be induced by PCBs (Rhee and Plapp, 1973). The profile of topically applied Aroclors is altered by this metabolism (Storr-Hansen et al., 1989) and metabolites are accumulated as well as excreted (Saghir and Hansen, 1993). Like many other invertebrates, insects lack the completely functional Ah receptor and, thus, are less susceptible than vertebrates to effects of the highly potent coplanar PCBs and PCDD/Fs (Hahn, 1998; Paine et al., 1993; Saghir and Hansen, 1995; Tehseen et al., 1992); therefore, because of slower penetration and excretion as well as reduced toxicities, they may contribute disproportionately higher TEQs to the predatory food chain.

4.8 Summary

The atmosphere contains a small total portion of global PCBs but is the major route of transport. The magnitude of differences between "contaminated" and "ambient" atmospheric PCBs is diminishing, but localized releases and reservoirs are retaining some distinctions. These distinctions are congener-specific, being closely related to congener properties such as vapor pressures, binding affinities and degradation potential. The more mobile lightly chlorinated and *ortho*-rich congeners are accumulating disproportionately in the colder and more remote regions of the globe. Biota, especially plants and insects, can capture vapor phase PCBs and introduce them into food chains.

5 THE HYDROSPHERE AND LITHOSPHERE

Transfer of PCBs as well as other persistent hydrophobic chemicals among environmental compartments is a continuous process. Biotic transfers transcend all major media. Atmospheric transfers of PCBs among aquatic, terrestrial and biotic reservoirs were introduced in Chapter 4. Thus, atmospheric transport is also important for this chapter. Chapters 4 and 5 are presented separately for convenience, even though the dynamic nature of mobile environmental PCB reservoirs makes compartmentalized discussions somewhat artificial. Hopefully, the information will be integrated adequately to reflect the actual continuum rather than the artificial segregation used for emphasis and clarity.

In many areas of the world, total mobile PCBs have declined significantly since the 1970s. Although some hot spots still exist and smaller recent releases are elevating ambient levels in certain regions, fish and human residues have generally declined in an encouraging manner. Some PCB degradation has contributed to this decline along with more controlled containment, but dispersion is probably the major contributor. "Today, the major source of ambient PCB exposure seems to be environmental cycling of PCBs previously released into the environment" (ATSDR, 1997).

"Fugacity" is a useful term for a generic summation of all the properties and forces contributing to the relative tendency of a chemical to escape from one phase and enter another. Although fugacity can be interpreted as a tendency to flee or become fugative, "fleeting" is too simplistic and "fugativity" is too long. Fugacity-at-work can be seen as regions formerly rather uncontaminated such as the Arctic and Antarctic, are accepting much of the dispersed mobile reservoirs because the lower temperatures lower the fugacities for otherwise mobile CBs (Wania and Mackay, 1993; 1996).

5.1. Relative Magnitude of Reservoirs

The proportions of total PCBs in different mobile environmental reservoirs were introduced in Chapter 4. Estimates of total PCB distribution among mobile environmental compartments used somewhat different categories. Both NAS (1979) and Tutsukawa and Tanabe (1984) considered the total ocean reservoir as mobile. Ocean water contains very low PCB concentrations, but the very high-volume of the open ocean resulted in estimated mobile loads as high as 66 x 10^6 kg in the US

North Atlantic and 230 x 10^6 kg in the open ocean worldwide (See Hansen, 1987b). Fresh waters may contain higher concentrations, but the dramatically lower total volume results in lower total amounts (**Table 5.1**).

Freshwater sediments also contain higher concentrations than do ocean sediments other than certain coastal hot spots; in this case, the net result is greater amounts of PCBs being found in freshwater sediments (**Table 5.1, Table 5.2**). By far, the greatest proportion of freshwater PCBs in the USA is found in The Great Lakes. World-wide proportions of total PCB loads are probably similar, with smaller scale deviations as in North America.

Tatsukawa and Tanabe (1984) estimated marine biota to contain greater proportions of PCBs than did NAS (1979); however, NAS estimated fresh water biota to contain a PCB load nearly equal to that in the freshwater itself (**Table 5.2**).

Each estimate in **Table 5.2** was based on similar total PCB data. Specific compartmentalizations and interpretations were different, but the estimates are adequately similar to serve as general concepts. The impressions provided permit a working understanding similar to the crude estimates of Aroclor production (**Tables 1.2** and **1.3**) and the relative proportions of congeners produced (**Table 2.6**).

Table 5.1. Total mobile PCBs in the USA and North Atlantic (NAS, 1979).

Compartment	PCB Load (10^6 kg)
ATMOSPHERE	0.018
HYDROSPHERE	
Freshwater	0.012 - 0.035
Sediments	1.4 - 7.1
Marine Water	6 - 66
Sediments	0.66 - 2.7
LITHOSPHERE	0.14 - 2.8

Table 5.2. Proportional estimates of total PCBs in various compartments.

Environmental Compartment	Percent Total	
	NAS[a]	World[b]
ATMOSPHERE	0.023	0.34
HYDROSPHERE		
Freshwater	0.044	
Sediments	9.02	
Biota	0.038	
Marine Water	83.86	99.5
Sediments	3.43	0.045
Biota	0.038	0.116
LITHOSPHERE	3.56	
Biota		

[a] NAS (1979) estimates ranged from 8.2 to 78.7 million kg total mobile PCBs (Table 5.1). Since the larger open ocean load is closer to other estimates, the higher ranges (total 78.7) were used for calculations.
[b] Marine environment only (Tatsukawa and Tanabe, 1984); atmosphere would average lower if remote land regions were also included. Ocean regions and biota were calculated as totals of more detailed estimates made by the authors.

5.1.1. Changing Characteristics of Reservoirs

It is well known that structures and properties of individual CBs influence selective biotic and chemical degradation; the individual properties also affect physical transfer among media. Thus, congener profiles in these declining residues are also changing. More highly chlorinated and *ortho*-poor CBs are less volatile, less water soluble and have a greater tendency toward particle binding and anaerobic dechlorination; thus, these congeners should be contributing relatively less to atmospheric and surface water profiles and more to residues in soils and sediments. The more lightly chlorinated and *ortho*-rich congeners are more rapidly degraded in mammalian systems, but are also more prevalent in the atmosphere, water surface layers, and fish from temperate waters.

The latter groups of CBs also appear to be increasing disproportionately in reservoirs and biota in the colder latitudes (**Section 4.4**). Anaerobic dechlorination favors *meta* and *para* positions, so product mixtures are not only more lightly chlorinated but are also proportionately *ortho*-enriched (**Section 5.2.3**). Frequently at undetectable levels in aged residues from mammals, the lightly chlorinated and *ortho*-rich congeners receive limited attention; however, they still contribute significant amounts to "static" reservoirs (i.e., unweathered former commercial mixtures), so that they also become transiently and locally important following accidental releases.

5.1.2. Global Models

Estimating total PCBs in global compartments was a useful starting point, but it was recognized that more complete analytical data were required. NAS (1979) concluded that further studies were "needed to firmly establish PCB concentrations and distribution in the marine environment." They also concluded that data bases were inadequate for modelling transfer among environmental compartments. It was further recommended that "EPA should continue to develop better analytical techniques.... Detailed information on individual isomers should also be reported together with the specific Aroclor designation." A few years ago, USEPA (1993) still recommended "Analysis of total PCBs, as the sum of Aroclor equivalents...." They did concede that "congener analysis is deemed to be a more scientifically sound and accurate method for determining total PCB concentrations. Consequently, States that currently do congener-specific PCB analyses should continue to do so."

The data base has expanded immensely since 1979, including a wealth of specific congener data thanks to improved analytical techniques and "under-focused" analysts. There have also been increased and more sophisticated efforts at modelling. Predictions have been improved, but assumptions are still necessary and the models do not always fit even slightly different situations. Models prefer first-order processes and equilibria. Failure to reach thermodynamic equilibrium in some organisms such as algae may be simply due to growth rates greater than uptake rates (Swackhamer and Skoglund, 1993). Nevertheless, the sum of all the transport processes superimposed on biotic transport and degradation and interacting with population fluxes and barely studied geophysical events defy the achievement of any stable equilibrium for PCBs in mobile reservoirs (e.g. Schreitmüller and Ballschmiter, 1994; Smith, 1995; Dachs et al., 1996).

In describing progression from Level I and Level II fugacity calculations (**Table 4.1**), Mackay et al. (1992) stated that "to assume equilibrium...is recognized as being excessively simplistic and even misleading." It is important to note that, once equilibrium is reached, there is no longer any net change. Steady state exchanges, on the other hand, need not be at equilibrium. If the exchanges among compartments approach reasonable steady states the models can be based on and detect changes in these nonequilibrium steady states. Morrison et al. (1996) found that a steady-state nonequilibrium model was superior to the equilibrium partitioning model for predicting bioaccumulation from sediments. Other factors favoring nonequilibrium distributions are discussed in **Section 5.2.2**.

The highly distinctive properties of individual CBs must also be recognized so that generic "chlorinated aromatic" models are used for guidance; accurate assessment of transfers and exposures requires more specific information. Even recent studies which distinguish congeners "*take mean values of too many measurements so that details for specific congeners may be obscured*" (Schreitmüller and Ballschmiter, 1994). Thus, a steady state for total PCB does not suggest the congener profile will be constant.

A reasonably linear relationship does not always mean the variations are merely chance. For example, **Figure 2.1** plotting GC retention times against B & Z number could easily be reduced to a straight line with a high correlation coefficient; however, the deviations from linearity are both real and predictable when the effects of chlorine substitution patterns are considered.

Still, global models are essential and detailed reliable baseline information must be established to assist their construction. For PCBs, the models must acknowledge non-uniform properties of individual CBs. Assumed equilibria may be necessary for initial models, but recognition of nonequilibrium conditions should govern their use. Steady states must also be recognized as congener-specific.

5.1.3. Global Models vs Micro-Reality

Each scientist has a different definition of the "big picture." Global models provide the concepts and information required as guidance for movement and exposure estimations, risk assessment, remediation options, and basic regulations. Each case, however, presents unique aspects which may be insignificant in a global-oriented conceptualization.

In the global distribution models, metabolism makes an infinitesimal contribution. For the individual organism, however, biotransformation can play a major role in both bioaccumulation and toxicities (**Chapter 6**). Components of food webs will also be affected directly by the biotransformation potential of fellow participants.

For persistent chemicals, existing residues are distributed over a larger volume as the animal grows. Thus, growth dilution can contribute to declines in concentrations without affecting total body burdens. The relative contribution of the "fat" compartment also increases in most species from birth to puberty and may show secondary bursts, especially with activity declines as in middle age. Food animals especially, due to confinement and high energy diets, are subject to major increases in size and relative increases in the proportion of body lipids. Growth dilution was found to reduce dieldrin residues adequately to make contaminated broiler chickens marketable (Combs et al., 1975). Growth and fat dilution of PCB residues has been readily demonstrated in swine (Platonow and Geisinger, 1973; Borchard et al., 1976; Hansen and Welborn, 1977). Redistribution of PCBs from peripheral to more central fat compartments has also been suggested (Hansen and Welborn, 1977; ATSDR, 1997). Confined laying hens have a tendency to develop fatty livers and, in spite of significant hexachlorobenzene elimination in egg yolks, fresh weight liver residues may increase with time since growth is insignificant (Hansen et al., 1978).

Growth dilution of persistent chemical residues is also common in aquatic species (Geyer et al., 1999). Changing composition of body lipids can also influence PCB accumulation by fish (Boon et al., 1984). Algae growth dilution exceeds uptake rate so that PCB equilibrium is not reached during (nonwinter) growth phases (Swackhamer and Skoglund, 1993). This dilution has a major impact on PCB residue concentrations (Camphens and Mackay, 1997). These authors presented a detailed model in which growth dilution played a major role in the net bioaccumulation of PCBs for several aquatic species. In an attempt to assess the relative contributions of major concentration-reducing pathways, various ratios were calculated from the model presented by these authors (**Table 5.3**).

It must be acknowledged that reducing a complex food web bioaccumulation model to a few ratios derived from information-packed diagrams is a gross oversimplification. Nevertheless, this monograph seeks to establish congener-specific relationships for individual organisms so comparisons of the relative contributions of different routes of gains and losses were of major interest. Concentration (or load) was expressed as mg/m^3 for consistency with other units (**Table 5.3**). This was species- and congener-specific and reported to agree within a factor of 3 with published data. The growth-loss (G_L) was a constant fraction of the load for each species, so was also appropriately species- and congener-specific (**Table 5.3**).

Table 5.3. Sample CB distributions and gains/losses[a] in a complex aquatic food web as predicted by a fugacity-based model (Camphens and Mackay, 1997).

CB	Mysid			Alewife			Salmon		
	Load mg/m^3	Gain (pg/da)		Load mg/m^3	Gain (pg/da)		Load mg/m^3	Gain (μg/da)	
		R_g	I_g/R_g		R_g	I_g/R_g		R_g	I_g/R_g
18	1.7	24.5	1	9.0	773	6	25.1	10.3	29
101	24.2	36.3	3	120.8	1157	30	419.7	15.5	269
194	3.0	0.8	10	9.7	24	76	23.0	0.3	587
	Loss (pg/da)			Loss (pg/da)			Loss (μg/da)		
	R_L/R_g	G_L	G_L/M_L	R_L/R_g	G_L	G_L/M_L	R_L/R_g	G_L	G_L/M_L
18	1.5	3.4	1700	4.7	1006	25	5.7	121	14
101	2.3	48.3	1610	6.6	13530	25	10.1	2023	14
194	1.3	5.9	no M_L	2.6	1087	25	2.7	113	14

[a]G = growth, I = ingestion, M = metabolism, R = respiration. Loss or gain designated by subscript L or G, respectively.

The ratio of G_L/M_L (Loss by metabolism) was logically species-specific but was modelled as congener-constant. Additional data were available, so that this ratio was approximately:

Alewife	25
Smelt	36
Sculpin*	36
Salmon	14

*The "3.6" calculated for CB 101 in the sculpin was probably due to a misplaced decimal.

Thus, loss by metabolism was 14-36 times slower than loss by growth, regardless of congener. The lack of congener-specific G_L/M_L ratios was alarming, but it must be recalled that biotransformation is much more important to the individual than to global PCB fluxes.

Still, CB 18 is of special interest since it may have been the highest production congener (**Table 2.6**). In minimally aged human as well as environmental residues, it makes a significant contribution, so biota are subject to pulses of exposure to CB

18 (Hansen, 1998). It has long been known that CB 18 is metabolized at least 10-times more rapidly than CB 101 by mammalian microsomes (**Chapter 6**) and within aquatic model systems (Metcalf et al., 1975). Recent studies, however, have demonstrated that more mono- and di-hydroxy metabolites can be detected for CB 101 than for CB 18 after extended microsomal incubations (Borlakoglu and Wilkins, 1993). This has been incorrectly interpreted as indicating that CB 101 is more rapidly metabolized because the secondary metabolites of CB 18 were not included (See **Chapter 6**).

Both CB 18 and CB 101 are primarily oxidized by CYPs 2B and 3A which are less important than CYPs 1A in fish; however, the rapid degradation of CB 18 in aquatic systems (Metcalf et al., 1975) and by insects and planarians (Saghir and Hansen, 1992; Saghir et al., 1993; 1994; 1995) suggests that there is potential for sporadic exposure to the parent compound and, probably, frequent exposure to metabolites (Hansen, 1998).

Other species/congener-specific relationships such as body burdens, ingestion:respiration gains (I_g/R_g), and respiration loss:gain (R_L/R_g) (**Table 5.3**) are as anticipated from congener occurrence and properties as well as biological principles. The evolution of very congener-specific models can be seen by comparing relative residues in **Table 5.3** with the global average fish concentrations used as crude estimates in earlier fugacity-based models (Mackay et al., 1992; **Table 4.1**). Fish were estimated to contain 26, 27, 28 and 28 pg/g CBs 18, 52, 101, and 128, respectively, in the early models and these differences can hardly be called congener specific.

Specific congener distributions are somewhat predictable, but more detailed relationships must be examined. For example, the weight percent of a "series" (progressive single chlorine additions) of tri- through hepta-CBs in various marine organisms indicates variations not readily predicted from global models (**Figure 5.1**).

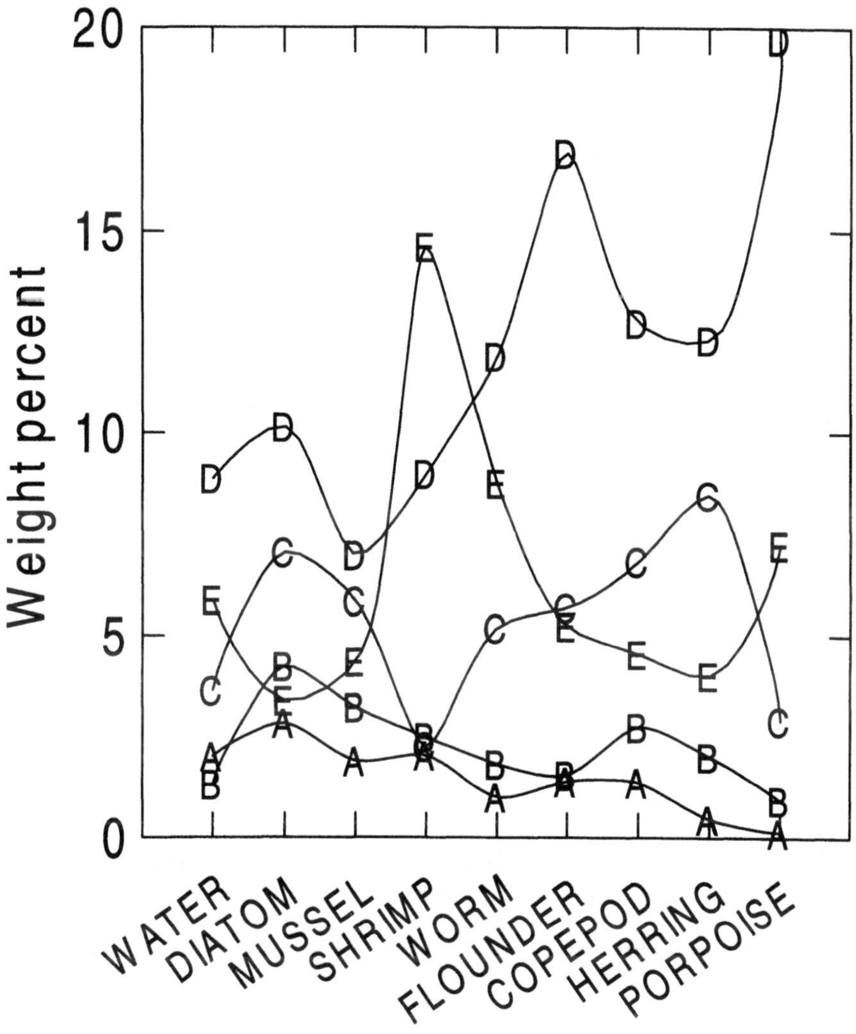

Figure 5.1. Weight percent of 25-2 (A), 25-25 (B), 245-25 (C), 245-245 (D), and 2345-245 (E) CBs in marine biota (Kannan et al., 1995). Mole percents would have changed proportions somewhat, but would also have decreased the resolution.

5.2. Movement and Fluxes of PCBs Among Reservoirs

Having differentiated between global concepts and local specifics, this section will provide general global estimates and proceed to specific congener values. Selected environmental PCB transfers will be presented such as **Table 5.4** to aid conceptualization. Equilibrium- and model-perturbing variables will be mentioned with regard to specific congeners. The goal is expansion and reinforcement of comparative concepts, not construction of mathematical models.

Table 5.4. Mean atmospheric PCB concentrations and fluxes in various locations (Eisenreich et al., 1981).

Location	Air (ng/m^3)	Precip (ng/L)	Flux ($\mu g/m^2/y$)
Urban	5-10	50	100 - 700
Rural	0.8	20	50 - 500
Great Lakes	1	20-50	20 - 150
Marine	0.5	1-5	0.1 - 20
Remote	0.1	5	0.2 - 20

5.2.1. Fluxes Between Water and Air

Atmospheric fluxes account for much of the PCB input into various aquatic and marine reservoirs (**Table 5.4**). These fluxes also maintain atmospheric concentrations via output. Inputs into The Great Lakes from atmospheric deposition (both wet and dry) have been estimated and revised, but the estimations by Eisenreich et al. (1981) still appear reasonable (**Table 5.5**).

Table 5.5. Total deposition of PCBs, DDTs, and polycyclic aromatic hydrocarbons (PAHs) into the Great Lakes (metric tons/year) (Eisenreich et al., 1981).

Lake	Total PCB	Total DDT	Total PAH
Superior	9.8	0.58	163
Michigan	6.9	0.40	114
Huron	7.2	0.43	118
Erie	3.1	0.19	51
Ontario	2.3	0.14	38

Improved sampling and analytical techniques have forced adjustments of some values. Moreover, total mobile PCB reservoirs have declined, so that the amounts in this table have been revised downward in more recent estimates and most sediments have become sources rather than sinks (e.g. Oliver et al., 1989; Pearson et al., 1996). Other roles also reverse as total input into mobile reservoirs declines. The atmosphere is now referred to as a "sink" for PCBs volatilizing from the Great Lakes (Hornbuckle et al., 1995).

For specific congeners, Henry's Law Constants (**Chapter 2**) govern relative transfers and net distributions between water and the overlying atmosphere. Some water solubilities and vapor pressures of individual chlorinated aromatics were presented in **Table 4.2** and the enrichment of the atmospheric profile in lightly chlorinated and *ortho*-rich CBs was established in **Chapter 4**.

Near-equilibrium CB profiles occur at the interface of the immediate overlying atmosphere and the surface of waters, but the profiles of the larger air and water reservoirs are influenced by a number of factors. The more volatile and mobile CBs disperse more rapidly in the atmosphere, while the heavier and coplanar CBs tend to be particle-bound and/or more readily degraded in the atmosphere (**Chapter 4**). In the water, the heavier congeners and less soluble congeners also are more likely to be bound and removed from the more freely exchanged state. Sample profiles are presented in **Table 5.6**.

In spite of dissimilar selected analytes, proportions of CB 110 (corrected) and CBs 138 + 163 are similar in both sets of air samples. CB 28 makes a decidedly lower contribution and CB 101 is decidedly higher in over-ocean air compared to over-lake air (**Table 5.6**). The high proportion of heavier congeners in the sample from the southern Atlantic is attributed to a profile characteristic of Trade Winds, which is discussed in detail (Schreitmüller and Ballschmiter, 1994).

Table 5.6. Sample relationship between surface water and air CB profiles compared with reported air CB profiles over the ocean.

	Lake Superior[a]		Atlantic Ocean[b]			
	Wt% of 35 peaks		Wt% of Total PCB			
Congener[c]	Water	Vapor	32N, 24W	2N, 29W	46S, 59W	16N, 23W
8	11.55					
16,32	3.38	11.21				
17	3.57	13.88				
18	5.34	**6.29**				
22	1.90	11.50				
28	5.47	11.29	3.11	3.21	0.68	2.02
31	7.40	**7.55**	4.95	5.77	1.09	2.74
33	4.31	3.95				
37,42	3.51	1.65				
41,64	1.91	1.21				
44	2.04	**2.35**	3.79	4.17	1.12	3.45
47,48	1.07	1.66				
49	1.83	2.10	3.16	2.88	0.65	2.26
52			11.16	11.22	3.04	8.21
60,56	2.31	1.50				
66,95[d]	5.07	3.25				
70	5.50	4.52				
74	1.94	1.30				
82[e]	0.46	3.87[e]				
87			1.32	1.28	1.64	1.55
97	0.73	**0.81**				
99	0.96	0.50				
101	4.31	1.29	7.74	6.73	7.38	7.62
110[e]	3.27	**0.24**[e]	2.26	2.56	3.77	3.10
118	3.46	1.29	0.79	0.96	2.34	1.55
138,163[d]	2.26	1.47	1.05	1.44	5.45	2.38

	Lake Superior[a]		Atlantic Ocean[b]			
	Wt% of 35 peaks		Wt% of Total PCB			
Congener[c]	Water	Vapor	32N, 24W	2N, 29W	46S, 59W	16N, 23W
141	0.85	**1.23**				
144	2.54	0.88				
146	2.61	0.19				
149			2.37	3.04	5.58	4.17
151			1.32	1.12	1.95	1.79
153	1.33	1.20	1.05	1.28	4.68	2.38
174	0.33	0.40				
175[d]	1.96	0.11				
180	3.65	0.34	0.11	0.32	1.43	2.38
185	1.56	0.39				
187	0.52	0.32				
196	0.49	0.11				
201[d]	0.57	0.12				
Total PCB	547 pg/L	1250 pg/m^3	190 pg/m^3	312 pg/m^3	385 pg/m^3	84 pg/m^3
% Total	100	100	44.16	45.99	40.80	45.60

[a] Baker and Eisenreich, 1990. Mean values for 5 samples from both phases. Shaded values were enriched in the vapor phase relative to surface water based on absolute concentrations as well as percent composition; bold values were near the approximate ratio of about 2 pg/m^3:1 pg/L for total PCB. Regular type indicates a ratio lower than that for total PCB.

[b] Schreitmüller and Ballschmiter, 1993. Four of 14 areas sampled; CBs as percent of total PCB which was estimated by multiplying the sum of 7 marker peaks (28, 52, 101, 118, 138, 153, 180) X 4.

[c] CB peak assignments from Baker and Eisenreich adjusted for current understandings.

[d] CB pairs 66/95 and 138/163 should be designated. CB 175 is usually a less significant component. CB 201 is currently referred to as CB 199 (2345-2356 octaCB); Appendix Tables 1 and 2.

[e] Nearly coeluting CBs 82 (rare) and 110 (high) may have been reversed in the quantitation of vapor PCBs (See **Section 3.2.1**).

5.2.2. Fluxes Between Water and Sediments

Although sediments have generally been considered sinks for semi-volatile compounds of low water solubilities such as PCBs, the relative roles of most sediments have reversed. As atmospheric and other inputs have declined, water concentrations decline and the large reservoirs in sediments are remobilized. This not only strengthens the argument against apparently unattainable PCB thermodynamic equilibria, but even makes achievement of steady states more elusive.

Even when input exceeded output, sediments were still dynamic reservoirs. Shallow sediments, of course, contribute PCBs to biotic burdens through bottom-dwelling organisms as well as maintaining a slow exchange with water. There is also considerable churning of shallow sediments, resulting in re-mobilization of bound as well as free PCBs. Dredging and transportation-generated churning can generate pulses of suspended particulates containing PCBs.

Direct recycling of the more water soluble congeners between the sediment and the water column plays a major role. Seasonally, fluxes of PAHs are much lower than those for PCBs and closely correlated with particulates and organic carbon (Sanders et al., 1996). In winter, PCB flux may be associated with sediment resuspension, but recycling events directly associated with the water are important during summer months. Sediment recycling accounted for about 2 to 20 times more PCBs in sediment traps than could be attributed to atmospheric deposition into Esthwaite Water (UK) (Sanders et al., 1996). Less chlorinated congeners accounted for most of the total PCB in sediment traps, with CB 18 + CB 28 contributing more than 20% to the total PCB profile.

Input/output budgets for Lake Michigan also showed large imbalances, indicating failure to understand all the processes of PCB transport to the water column (Pearson et al., 1996). Water column recycling of more lightly chlorinated (and *ortho*-rich) congeners from sediments is a direct path to atmospheric PCBs (e.g. Chapter 4; **Table 5.6**; Baker and Eisenreich, 1990; Sanders et al., 1996). This pathway favors more lightly chlorinated and *ortho*-rich CBs because of greater water solubility and volatility and lesser trapping by particulates and colloids. In spite of this, the PCB profile shifted toward more lightly chlorinated homologs (diCBs through pentaCBs). Consequently, these congeners had half-lives greater than that for total PCB while the more heavily chlorinated congeners had shorter residence times in the open water (Pearson et al., 1996). One of the input processes not considered was anaerobic dechlorination (**Section 5.2.3**) Although slow, this process would exaggerate profile changes by introducing more lightly chlorinated CBs at the expense of, e.g., hexaCBs.

Surface water PCBs essentially exist in 3 phases - dissolved, bound and colloid associated (e.g., Baker and Eisenreich, 1990); sediments are in the same 3-phase equilibrium with pore water (e.g. Burgess et al., 1996). In both cases, a greater proportion of the more water soluble lower chlorinated (and *ortho*-rich) congeners are in the truly dissolved phase. In sediments, pore-water PCBs are less than 0.05% particle bound residues; nevertheless, the congener-specific movements toward equilibria could make this a significant distinction, further favoring aqueous solvation of the more water soluble CBs. **Table 5.7** illustrates the homolog profiles of shallow (approximate low tide = 1 m) sediment cores from New Bedford Harbor compared to the profiles of the major Aroclors contaminating the harbor (Burgess et al., 1996).

Table 5.7. Homolog composition and interstitial water:colloid distribution of PCBs in shallow New Bedford Harbor sediments.

Homolog	Percent Homolog Composition			Ratio of Coll/Free
	A 1242	A 1254	Sediment	
DiCB	14.7	< 0.1	28.7	1.2
TriCB	46.0	1.8	29.4	2.9
TetraCB	30.6	17.1	21.9	4.7
PentaCB	8.7	49.3	13.1	5.8
HexaCB	< 0.3	27.8	5.6	6.1
HeptaCB		3.9	0.7	6.1
OctaCB		< 0.5	0.5	8.0

The colloid:free ratios increase with increasing homolog group, but specific congeners within these groups also vary according to solubilities and sediment content. Approximate profiles of the major congeners found in highly- and moderately-contaminated upper New Bedford Harbor sediments are presented in **Table 5.8**. These profiles along with presumed water:air steady state profiles in **Table 5.6** suggest that some *ortho*-rich tetra-, penta- and even hexa-CBs partition similarly to some more lightly chlorinated *para*-rich congeners.

Table 5.8 Percent distribution of major congeners (4% or greater of at least one sample type) in sediment, water and ribbed mussels in two areas of upper New Bedford Harbor (values estimated from figures *in* Lake et al., 1995).

B & Z No	High Contamination				Moderate Contamination			
	Mussel	Sed	Water	W/S	Mussel	Sed	Water	W/S
8	1.9	5.7	7.1	1.2	1.1	1.5	4.4	2.9
18	2.5	3.8	4.5	1.2	2.0	1.8	4.0	2.2
26	5.0	4.8	6.5	1.4	3.9	1.5	5.3	3.5
28	6.3	4.9	5.1	1.0	5.8	4.1	6.0	1.5
31	6.7	6.4	8.0	1.3	6.2	6.0	7.1	1.2
47+48	4.7	4.5	3.6	0.8	4.8	3.8	3.8	1.0
49	5.4	5.1	4.7	0.9	5.5	4.1	4.7	1.1
52	6.1	5.6	6.0	1.1	6.1	4.4	5.4	1.2
70+76	3.4	2.0	2.0	1.0	4.5	6.1	2.8	0.5
95+66	6.4	5.4	4.9	0.9	6.9	8.5	5.2	0.6
110	5.6	4.8	4.0	0.8	6.1	5.2	4.3	0.8
153	3.1	3.6	2.2	0.6	4.0	3.4	2.8	0.8
Total ng/kg	732	2330	351		128	1870	20	
% Total	57.1	56.6	58.6		56.9	50.4	55.8	

Sediments also contain large amounts of chitin, the polymer composing arthropod exoskeletons. This aminopolysaccharide adsorbs PCBs selectively from the water; it also acts as a substrate for bacteria so chitin may facilitate dechlorination by bringing the components together (Next Section). Other microscale influences include facilitated solvation of PCBs by water miscible solvents such as acetone and ethanol (Li and Andren, 1995); conversely, PCB "solventing out" of the water by partially miscible solvents such as chloroform and trichloroethylene is also possible (Coyle et al., 1997).

Sediments exposed directly to the atmosphere during water level changes may rapidly transfer volatile congeners directly to the air through hydration-enhanced mechanisms (Chiarenzelli et al., 1996; 1997). Once dehydrated, binding is tighter and net volatilization diminishes (**Chapter 4**).

Even deeper sediments such as Lake Erie and Hamilton Harbor may transport PCBs from sediments directly to the atmosphere through ebullition of gas (mostly nitrogen and methane) bubbles (Adams et al., 1990). The Henry's Law Constants of PCBs (between negative log 3 and 5) are in the range where transfers between air and water are about equal, but the higher values encounter more resistance entering the gas phase (See Eisenreich et al., 1981). The authors suggest significant transport of lower chlorinated homologs to the surface by bubble ebullition, but *ortho*-rich pentaCBs 88 through 96 and CB 136 have lower Henry's Law Constants than diCB 15, triCBs 22 and 37, and especially tetraCBs 77 and 81 (**Appendix Table 3**); thus, *ortho*-rich CBs should also be mobilized by this pathway.

Vertical fluxes of organic matter (e.g. plankton, fecal pellets) and associated PAHs and PCBs to and from the still deeper sediments of the Mediterranean have been shown to be governed by factors contributing to "flicker noise" (Dachs et al., 1996). Both sedimentation and upwelling are governed by population dynamics as well as geophysical phenomena. The authors suggest there are area-dependent self-organizing principle(s) which may seriously compromise conclusions based on short-term or infrequent samples. Presumed PCB sinks in deep ocean sediments may also be re-mobilized periodically. Tremendous bottom turbulence is created in the western basins of the Atlantic Ocean when cold Antarctic bottom waters attempt to pass through gaps in fracture zones (Polzin et al., 1996).

These considerations alone indicate that the hydrosphere is in a constant state between order and chaos; consequently, it is doubtful that subcompartments and biota, even in smaller hydrosystems, ever approach stable equilibria (see Section 5.1.2.) and steady state transfers will be sporadically perturbed. This also affects atmospheric equilibria because of the major fluxes between water and air (Section 5.2.1).

The time scale required for troposphere/oceanic surface water equilibrium for anthropogenic compounds such as PCBs is not known; steady state transfers are unique to local conditions. Distinctly different CB patterns were observed at the ocean/air interface in the Atlantic and correlated with numerous variables illustrating subtle factors and major differences in PCB congener behaviors (Schreitmüller and Ballschmiter, 1994) (**Table 5.6**). Some *ortho*-rich tetra- (CB 52), penta- (CB 101) and hexa- (CB 149) chlorinated congeners make larger contributions to the over-ocean atmosphere than do triCBs 28 and 31, especially in specific conditions.

Solvation of PCBs from sediments by water, further driven by vaporization (both from the surface and via ebullition), should continuously deplete sediments of *ortho*-rich and lightly chlorinated CBs. **Table 5.7** suggests, however, that the opposite might be true. The relative proportions of more lightly chlorinated CBs increase relative to the proportions in the Aroclors which were the source(s) of the residues. Influx of these more mobile CBs must also be occurring at rates exceeding their efflux. This is also reflected in the longer half-lives of lower chlorinated homologs compared to higher chlorinated homologs in the open water of Lake Michigan (Pearson et al., 1996). Air/water steady-state exchanges may partially explain the persistence of lightly chlorinated homologs in the water column, but the sediments should move toward depletion of these CBs. Mass balance calculations would suffer if losses of higher chlorinated homologs resulted in gains of more lightly chlorinated CBs. One major source of lightly chlorinated and *ortho*-rich CBs is the anaerobic dechlorination of higher chlorinated congeners.

5.2.3. Anaerobic Dechlorination in Sediments

Oxidative metabolism of PCBs is generally considered a degradation pathway, selectively removing the more biotic-labile congeners from reservoirs. Oxidative metabolism and metabolites are discussed in **Chapter 6**. Anaerobic dechlorination by adapted microbes has a more profound influence on PCB profiles since the products are also PCBs. The profile changes are greater than simple depletion because other congeners are produced from the dechlorinated CBs. Although microbial dechlorination is a relatively slow process, it occurs most readily in highly contaminated sediments and in a few years can alter profiles adequately so that the parent Aroclor may be mistaken (e.g. Bedard and May, 1996; **Table 5.7**).

Short-term profile changes in localized pockets may not be detectable because of the other forces favoring selective removal of the products relative to the parent CBs (previous section; Hansen, 1987b). Even long term changes in, e.g., Lake Ontario (Oliver et al., 1989) may not be obvious because of selective removal of products.

Anaerobic dechlorination of PCBs has only been studied intensively for about a decade (e.g. Brown et al., 1987; Quensen et al., 1988). Several outstanding reviews and in-depth papers are already available, attesting to the perceived importance of this semi-degradative pathway (e.g. Abramowicz, 1990; Abramowicz et al., 1993; Bedard and May, 1996; Brown et al., 1990; Kim and Rhee, 1997; Quensen et al., 1990; Rhee et al., 1993; Tiedje et al., 1993).

An important facilitator of the process, chitin, is an enormous product from aquatic and marine invertebrates (Portier and Fujisaki, 1988; Øfjord et al., 1994). Detectable dechlorination requires relatively high PCB concentrations, so enrichment in both PCBs and microbes should favor the process. Sedimented chitin can carry and/or adsorb PCBs as well as serve as a growth substrate for sediment microbes. The adsorption is CB-specific, so profiles might be distorted if extractions are not complete.

When pattern changes are discernable, specific patterns are identifiable for specific sources of microbes. Nearly all pathways involve *meta*- and *para*-dechlorinations so that the resulting profile is *ortho*-enriched. Anaerobic dechlorination of common highly chlorinated congeners is a source for normally rare substitution patterns such as 2,4,6-CBs (Bedard and May, 1996; **Appendix Table 9**). Although these CBs do not reach dominant proportions in the mixes, proportions of other significant congeners such as CBs 28, 47, 48 and 49 may be greatly increased (Bedard and May, 1996; Øfjord et al., 1994; Rhee et al., 1993). Mono-, di- and trichloroCBs produced are generally considerably more volatile and water soluble, resulting in both water and air enrichments; however, because these congeners are also more mobile, the enrichments may be transient.

Thus, PCB-contaminated aquatic and marine sediments/waters and even perpetually moist soils may develop residue profiles that are not readily recognized unless this important process is considered. The changing composition of the residues alters both distribution and toxicity characteristics.

5.2.4. Fluxes Among Water, Air and Soils

Sediments and soils received net PCB inputs from water and air during the peak PCB emissions of the 1960s-early 1970s. Similar to the reversed roles of sediments from sink-to-source, soils now appear to be a reservoir for releasing PCBs into the atmosphere. Since the atmosphere now can be considered a net recipient of PCBs from soil, water and (indirect) sediment outputs, all reservoirs are still interconnected.

The 209 CBs span a broad range of volatilities. Lightly chlorinated and *ortho*-rich congeners readily move from soils and waters in warmer latitudes; colder temperatures favor partitioning back out of the atmosphere, retard re-volatilization, and slow decomposition reactions (Wania and Mackay, 1996). The more heavily chlorinated and *ortho*-poor/*para*-rich CBs require more energy to be volatilized and are more readily "condensed" from the atmosphere; thus, they are considerably less mobile.

Table 4.1 has summarized selected examples of modelled fugacity-based environmental distributions (Mackay et al. 1992). As anthropogenic releases declined, the roles of the reservoirs reversed and some modification of steady state assumptions are required.

The relative soil accumulation and soil degassing of 4 congeners (CBs 28, 52, 138 and 153) were estimated for the period 1942 through 1992 (Harner et al., 1995). Based on a 2-compartment model, a half-time of 10 to 20 years was estimated for soil recovery from the high pulse of input between 1960 and 1970. The more mobile CBs 28 and 52 should approach soil:air equilibria by about year 2000; however, CBs 138 and 153 are much slower to volatilize and will continue to be degassed from the soil for a longer time. Although considered an adequate accounting, the authors suggest several improvements could be achieved by considering additional factors including vegetation buffering.

Biota have a minor global impact, but should have a major local impact on PCB profiles. Plants are known to accumulate vapor phase PCBs (**Chapter 4; Appendix Table 7**). Large differences in soil congener losses have been shown to correlate better with vegetation cover than with the physical constants of individual CBs (Grundy et al., 1996).

Atmospheric congener profiles over PCB-contaminated landfills tend to be distinct from atmospheric profiles over rivers, lakes and oceans (e.g. **Chapter 4; Appendix Table 6**; Jan et al., 1988; Hornbuckle et al., 1993; Chiarenzelli et al., 1996; Hansen et al., 1997). The presence of water further alters the net movement of specific CBs because of the soil ↔ air ↔ water exchanges. Biota are also known to alter water profiles of PCBs, facilitating and/or retarding the movement and degradation of specific congeners (e.g., **Table 4.1, Figure 5.1**).

5.3. Bioconcentration of PCBs from Water

Lipophilic PCBs readily cross membranes and attain higher concentrations in lipids than in aqueous media. The term "bioconcentration" is reserved for the process of extracting lipophiles from the water by biological lipids. Bioaccumulation refers to the net retention of chemicals in an organism after net input and output through various routes reach steady states. The progressive increase in biotic residues at higher trophic levels is referred to as biomagnification. Specific ratios (factors) are referred to as BCFs, BAFs and BMFs, respectively.

Bioconcentration is a useful parameter for comparing different lipophilic chemicals such as PCBs. As would be expected, BCFs are directly proportional to partition coefficients and lipid contents (ATSDR, 1997; Geyer et al., 1999). Determination of accurate BCFs require flow-through exposures below the limits of solubilities for long periods of time to establish steady states (Geyer et al., 1999).

BCFs in various aquatic and marine organisms are generally in the range of 500 - 40,000 for lower chlorinated PCB congeners and about 11,000 - 300,000 for CBs 70, 101, 110 and 136 (Geyer et al., 1999). BCFs determined or calculated on a lipid basis range from 11,000 to 10,000,000 for the same congeners. Coplanar CBs have higher BCFs and the more highly chlorinated congeners may have organism-based BCFs as high as 2,000,000 and lipid-based BCFs around 20,000,000.

Because BCFs consider a single transfer between phases and must be corrected for dilution by growth, elimination and degradation, they are only one of several factors contributing to net bioaccumulation. If one is concerned about the health of aquatic and marine environments and/or the dietary intake of PCBs from fish and seafood, BCFs alone will not accurately predict the profile of CBs encountered. The other forces are at least as congener-specific as is water-lipid bioconcentration.

Although a direct relationship between partition coefficients and bioaccumulation would make predictions simpler and more accurate, several authors have shown that this can be a misleading oversimplification. Lipid-based PCB sediment/fish BAFs are higher and more homolog-dependent in fish from nonpolluted lakes than in fish exposed to high sediment PCB concentrations (van der Oost et al., 1994). For sediments in Lake Erie (with comparable high PCB concentrations), sediment/invertebrate BAFs were found to more closely resemble a parabolic than linear relationship to partition coefficients (Morrison et al., 1996). Curvilinear relationships between congener partition coefficients and even BCFs have also been demonstrated (Fox et al., 1994). Invertebrate/sediment BAFs and tree swallow/sediment BMFs were predictable based on fugacity theory for coplanar and mono-*ortho* CBs when normalized to sediment organic carbon (Froese et al., 1998b). Total PCB BAFs and BMFs were 9- to 30-fold higher attesting to the high fugacities (low binding) of nonplanar CBs.

Brown (1992) considered alterations in specific residue profiles to be indicative of metabolism, but in some species, such as striped bass from the Hudson River, prey selection played a bigger role than metabolism or BCFs. Prey-to-predator transfers of total PCB were considered the principal source for Great Lakes salmonids and appeared to be constant in data obtained after 1983 (Jackson and Schindler, 1996). On a specific congener basis, however, Brown (1992) and Kannan et al. (1995) have demonstrated that metabolism plays a major role in residue profiles and, thus, net BAFs.

5.3.1. Metabolic Slopes

BCFs for PCBs are congener-specific, but dilution by growth should be a constant factor for the entire PCB profile. Biotransformation or metabolism makes a smaller contribution to total PCB residues and an even smaller contribution in global models. Nevertheless, metabolism is perhaps the most congener-specific of the processes affecting net bioaccumulation. Metabolism is also highly species-specific and influenced to varying degrees by disease, nutrition, and the presence of other chemicals as well as the profile of PCB residues. **Chapter 6** presents a more complete discussion of metabolism and **Chapter 8** incorporates many of these variables into discussions of food chains.

These relationships help to explain certain species (and higher taxa) differences in accumulated and retained PCB residue profiles. For example, fish from temperate waters generally have lower basal levels and less inducible forms of PB-type CYPs than do birds and mammals; thus, they tend to retain CBs which are normally quite labile and frequently undetected or (unreported) in mammals. This is *not* an absolute distinction. Such generalizations (Stegeman and Lech, 1991) are often over-interpreted and high basal as well as inducible CYP 2B-like and CYP 3A-like proteins have been identified in several species, especially of tropical fish (Stegeman et al., 1997). Furthermore, the anticipated CYP 1A-like activities are not universal in fish species and may be suppressed rather than (or as well as) induced by coplanar CBs (White et al., 1997).

Fish residue profiles are still unique and influence the residues, if only sporadic, in warm-blooded fish eaters. Heidmann and Beyerbach (1989) compared residues in eggs of terns and oyster catchers directly to a 1:1 mixture of Aroclor 1254 and Clophen A60 to determine metabolism parameters. Although their method detected some species and congener differences, they acknowledged it was a preliminary study. With the absence of intervening profile changes, this method actually estimates the summation of all degradative and advective processes within the ecosystem rather than just metabolic capacity of the birds.

Other authors used biotic residues to determine food chain relationships (e.g. Boon and Duinker, 1986; Focardi et al., 1988; Muir et al., 1988; Norstrom et al., 1988). Later studies also used comparative environmental residue profiles to develop specific congener relationships (e.g. Brown, 1992; Froese et al., 1998b; Kannan et al., 1995; Koslowski et al., 1994; Muir et al., 1995; Nichols et al., 1995; Pruell et al., 1993; Tillett et al., 1996).

Various schemes have been devised to classify CBs according to metabolic turnover. In this section, emphasis will be placed on aquatic/marine (especially fish) residues in preparation for expanded discussion of food chain transfer patterns in **Chapter 8.**

5.4. Bioaccumulation from Multiple Pathway Exposures

Kucklick et al. (1994) found a direct (metabolism-independent) relationship between CB partition coefficients and BAF patterns in fish and seals from Lake Baikal (Siberia). The correlation plots resembled linearity (as in **Figure 2.1**), but several factors could have obscured any effect of differential metabolism. Dramatic losses of CBs 8, 17/18, 16/32, and 52 between the fish and the seal profiles were not acknowledged. CB 153 was accumulated more than 2-fold by the seal compared to fish, even though the peak also contained CB 105 and the labile CB 132. Other anticipated decreases in relative contribution (e.g. CB 110) occurred, but CB 101 contributed a nearly equal proportion of sculpin, predatory omul and Baikal seal residues. Other selective distributions that might have been overlooked because persistent and labile congeners co-eluted included CBs 66/95, 118/149 and 138/163.

Kannan et al. (1996) concluded that metabolism perturbed the direct relationship between log K_{ow} and BAFs. The raw data for total PCBs and congener percentages are presented in **Table 5.9**. A larger number of organisms and better congener resolution permitted normalizations of data:

1) congener concentrations were converted to molar fractions of CB 153 concentration ("unmetabolized");

2) these ratios were plotted against the molar/153 ratio for the diatom which was considered to have no biotransformation potential.

Table 5.9. Weight % composition of PCB residues in organisms from a marine food chain[a] based on ng/g of lipid or, for water, pg/dm^3 (Kannan et al., 1995). Copepod (herring prey) and sandworms (flounder prey) were not included in the table below, but were reported to segregate by principal components analysis with their respective predators (**Figure 5.1**).

B&Z No.	Water	Diatom	Mussel	Shrimp	Flounder	Herring	Porpoise
18	2.0	2.8	1.9	2.0	1.4	0.5	0.1
44	1.1	3.5	2.3	2.0	2.6	3.0	1.0
49	6.4	2.3	1.3	1.6	1.1	0.8	0.5
52	1.3	4.2	3.2	2.5	1.5	2.0	0.9
70	1.2	2.7	2.3	1.1	1.1	2.2	0.3
74	0.4	1.0	1.1	0.7	0.5	0.9	0.6
84	0.2	1.7	1.5	0.6	0.6	0.8	0.2
87	1.8	2.5	2.1	1.2	1.1	2.2	0.6
91	0.3	0.6	1.1	0.7	0.5	0.7	0.3
92	1.6	1.3	1.5	0.8	1.1	2.2	0.9
97	1.0	1.4	0.6	0.3	1.7	1.3	0.2
99	1.1	2.1	1.3	1.1	2.0	2.8	2.6
101	3.6	7.0	5.8	2.2	5.7	8.5	2.9
105	3.6	1.1	1.9	0.6	0.8	1.3	0.6
110	6.9	6.8	11.1	1.7	3.2	3.8	0.8

B&Z No.	Water	Diatom	Mussel	Shrimp	Flounder	Herring	Porpoise
118	5.8	4.9	4.0	3.4	3.7	4.7	2.6
128	2.3	1.5	1.7	0.3	1.2	1.6	0.1
132	3.4	3.9	5.3	2.1	3.1	2.1	3.2
138	21.9	12.5	14.0	15.7	16.9	16.2	21.1
146	0.8	1.7	1.5	7.1	2.9	2.5	2.2
149	4.0	6.5	4.7	2.1	7.1	7.1	6.1
153	8.9	10.1	7.0	9.0	16.9	12.3	19.7
156	3.6	1.5	1.3	3.1	0.6	0.9	0.0
170	1.6	1.7	3.2	3.7	1.8	1.3	2.2
172	0.4	0.6	0.9	1.6	0.3	0.4	0.4
174	1.6	2.4	3.8	2.6	2.6	2.2	2.5
180	5.9	3.4	4.3	14.6	5.2	4.0	7.2
183	0.8	1.1	0.6	2.0	1.8	1.3	2.4
187	1.3	1.4	2.6	7.2	3.7	2.9	4.9
194	1.2	0.3	1.1	0.7	0.3	0.2	0.3
SUM	16	710	530	890	650	1300	7600

[a] All samples collected in the southern Baltic Sea at 54°22'N, 10°00' E in March 1990.

The slopes for each of the four groups of CB congeners compared to the diatom separated the organisms by metabolism potential and the CB groups by metabolism. CBs were divided into 4 groups based on vicinyl hydrogens which is predictive of susceptibility to CYP 1A-type and/or CYP 2B-type oxidative metabolism. A separate paper describing similar relationships for 3 porpoises generally confirmed the slopes (Bruhn et al., 1995). Looser and Ballschmiter (1998) found similar relationships between freshwater prey and predator fishes.

The classifications were consistent with factors known to affect biotransformation (**Chapter 6**) and those of Brown (1992). Some irregularities were noted which may be due to variability (**Figure 5.1**, presented previously), but the general separations are logical. Still, the actual relationships deviate from predictions at a frequency that does not permit total dependence on these models (**Figures 5.2 and 5.3**).

It is apparent that some reconsiderations of PCB congener "types" is necessary. The vicinyl hydrogen classification does not consider arene oxide formation between Cl- and H-containing carbons leading to NIH Shift products (**Chapter 6**). Coplanar CB 77 is especially vulnerable to *para*-hydroxylation via this pathway. CB 110 has both vicinyl (3,4; 5',6') hydrogens and vicinyl (3',4') chlorines and the dual susceptibility to metabolism makes this major congener generally labile.

The increased proportion of CB 110 in mussels and drastic reduction in shrimp (**Figure 5.2; Table 5.9**) is consistent with the lack of metabolism potential suggested for molluscs and mixed metabolism potential for arthropods (Brown, 1992). It is also consistent with comparative 2,3,6-pentaCB (84, 95, and 110) profiles in mussels and shrimp reported by Lake et al. (1995). CBs 52, 101 and 151 were modestly metabolized by sandworms and dramatically metabolized by grass shrimp as compared to clams (Pruell et al., 1993). The shrimp also accumulatedrelatively less CB 138 (+163, which has vicinyl 3',4'-Cls and 5',6'-Hs) and 128 than did clams and sandworms, but relatively more CB 180 (Pruell et al., 1993). On the other hand, Morrison et al. (1996) reported practically equal proportions of five similar pentaCBs in freshwater mussels and crustaceans except that CB 105 and CB 110 proportions were lower in an insect and crayfish than in mussels and gammarus (**Figure 5.3; Appendix Table 9**).

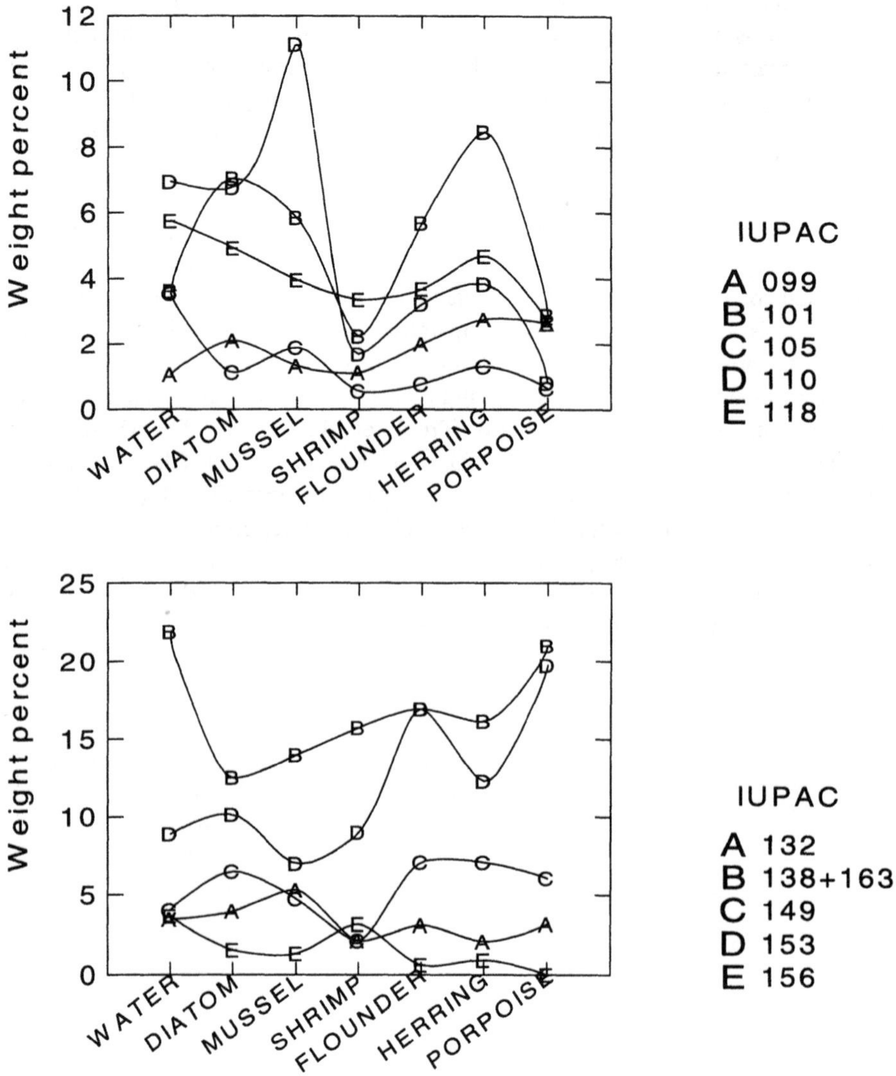

Figure 5.2. Marine food chain relationships of individual pentaCBs (top) and hexaCBs (bottom) from **Table 5.9** (Kannan et al., 1995).

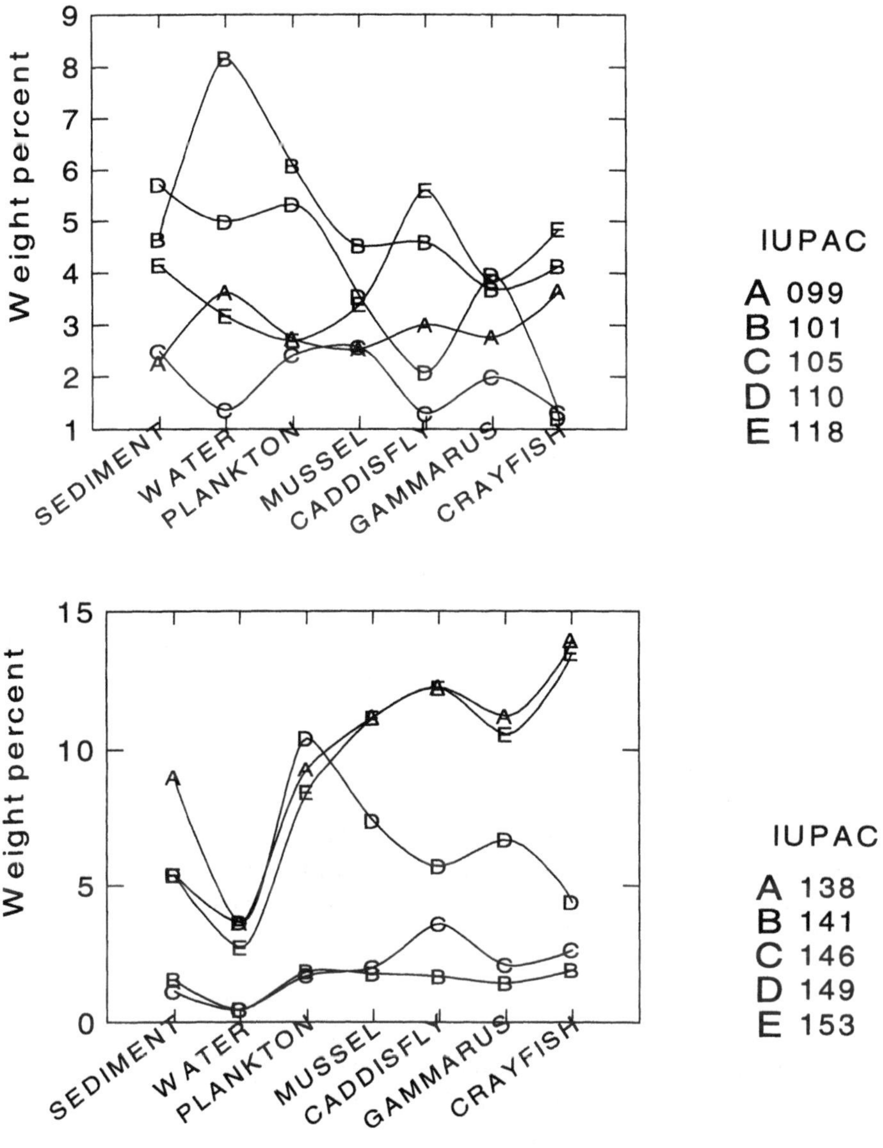

Figure 5.3. Aquatic food chain relationships of individual pentaCBs (top) and hexaCBs (bottom) from **Appendix Table 9** (Morrison et al., 1996).

The separation of CBs 138/163 was not addressed by any of the groups. This persistent/labile CB pair was the major peak in mussels, crabs, mullet and tuna from the Western Mediterranean (Porte and Albaigés, 1993). CB 163 was less than 10% of the peak in dead harbor seal pups (Troisi and Mason, 1997); however, CB 163 can contribute to 30% or more of the peak in other human and environmental samples (Larsen et al., 1994; Vetter et al. 1995).

Several authors resolved CBs 132 and 153. For hexaCBs, CB pairs 132/153 and 138/163 were parallel in the aquatic food chain (**Figure 5.3**); the divergent patterns for 138 and 153 in the marine food chain (**Figure 5.2**) may reflect the removal of the CB 132 influence on reported CB 153 proportions. CB 132 contributed about 10-30% of the peak in various fish, porpoise and seal samples (Bruhn et al., 1996; Kannan et al., 1996; Looser and Ballschmiter, 1998; Troisi and Mason, 1997), but contributed only about 2% to the peak in the blubber of a seal from the Caspian Sea (Vetter et al., 1995).

Morrison et al. did not consider metabolic transformation to be important in their model, but the clear divergence of the labile CB 149 from the distribution patterns of the other hexaCBs (**Figure 5.3**) suggests a selective metabolism influence. CBs 132 and 149 each contributed about 4% of the total PCB in both blubber and liver from dead seal pups which had higher CYP 2B than 1A activity (Troisi and Mason, 1997); CBs with vicinyl Cls (110 and 163) were 5 and 2% in blubber, but less than 1% in the livers of these pups. CBs 101, 149, 146, and 163 exceeded the concentration of CB 118 in blubber of a Caspian Seal (Vetter et al., 1995).

In quail fed Aroclor 1260 for 25 days, the CB 149 (95%) + CB 118 (5%) peak was accumulated above Aroclor proportions in livers of birds receiving 3 mg/kg Aroclor 1260 in the diet (Marsili et al., 1996). At higher doses which resulted in enzyme induction, the 149+118 peak (as well as other labile CBs) was greatly reduced in the serum and especially the liver. The ratio of CB 149/CB 153 has been suggested as an indicator of the level of biodegradation of PCB residues in the food chain (Turrio-Baldassarri et al., 1997).

Phytoplankton are generally considered merely intermediaries in PCB transfers; however, PCBs can alter the population profiles of algae (See Hansen, 1987a; 1987b) and plants can selectively metabolize certain CBs (Moza et al., 1973; 1979; Wilkin et al., 1995). Only CBs 99 and 101 deviate from the bioconcentration trend by the mixed plankton of **Figure 5.3**, while the diatoms in **Figure 5.2** show no bioconcentration of pentaCBs and apparent metabolism or exclusion of CBs 138/163.

Freshwater snails are able to *O*-dealkylate *p*-nitrophenol ethers and methoxychlor *in vivo* at rates comparable to those of fish (Hansen et al., 1972). Subcellular fractions of bivalve organs are able to oxidatively metabolize benzpyrene, but probably by peroxidative as well as monooxygenase pathways (Stegeman, 1985). The *in vivo* significance of these pathways at environmental levels is unknown, but molluscs may well have PCB metabolic capacities with SARs totally distinct from other phyla.

5.4.1. PCB Residue Profiles in Fish

The previous discussion of aquatic and marine food chains may have seemed detailed and tedious. In fact, it barely covered a small portion of the data and models available, but should have established a sense of dose-, congener-, species- and time-dependent PCB profiles. Additional inferences regarding invertebrate CYPs can be extracted from Hahn's (1998) review on the phylogeny of the AhR.

Fish PCB residue profiles are appropriately considered to be unique from those of other vertebrate classes. They are also unique among the various pisces taxa. Brown (1992) and Stegeman (1997) have demonstrated differences in PCB oxidative metabolism potential among various fish species. **Table 5.9** includes CB profiles from two marine fish and **Appendix Table 10** presents some specific congener profiles in Great Lakes fish which further attest to species specificity.

CYP P450 oxidative metabolism is only one of many determinants of net residue profiles. Relative solubilities, binding and gain/loss ratios (**Table 5.3**) for individual CBs are important. For fish, relative position in the water column interacts with CB properties so that pelagic tuna tend to bioconcentrate *ortho*-rich pentaCBs 84, 95, and 110 in muscle to proportions equivalent to the "major" CBs 153, 170, 174 and 180 (Porte and Albaigés, 1993). CB 118 was higher than CB 153 in tuna muscle and CB 138 (+163) dominated. The benthic mullet from the same areas of the Western Mediterranean accumulated relatively smaller proportions of CBs 84, 95 and 110, but CBs 118 and 149 were equivalent. In both fish, CBs 153, 180 and 187 constituted a greater proportion of the liver residue (relative to CB 138) than of the muscle residue.

Seasonal factors such as diet, migration, growth and/or fat mobilization, and spawning can apparently have large influences on residue disposition: in tuna collected in July, liver total PCB concentration was less than half muscle total PCB; tuna collected in October had 22 times higher PCB concentrations in liver than in muscle (Porte and Albaigés, 1993).

In general, fish tissues are enriched in the mammalian-labile congeners 52, 84, 95, 101, 110, 132, 141, 146, 149, 151, and 163 compared to the major congeners most frequently reported (References in **Appendix Table 10**; **Table 5.9**; Boon and Duinker, 1986; de Boer et al., 1994; Kucklick et al., 1994; Looser and Ballschmiter, 1998; Lake et al., 1995; Maack and Sonzongni, 1988; Macdonald et al., 1993; Muir et al., 1988; Newsome and Andrews, 1993; Porte and Albaigés, 1993; Schantz et al., 1993; Viganó et al., 1994; etc.). Lower chlorinated CBs and dechlorination products such as CBs 47 and 49 may dominate in some cases. Many of these congeners continue to make significant contributions to residue profiles in marine mammals and humans who consume fish on a regular basis (e.g. Duarte-Davidson and Jones, 1994; Green et al., 1996; Kannan et al., 1996; Lake et al., 1995b; Turrio-Baldassarri et al., 1998; Vetter et al., 1995; Zuccato et al., 1999).

Some seemingly unique distributions may be attributable to the selective partitioning and/or metabolism of optical isomers of *ortho*-rich congeners. Atropisomers of several of these significant congeners are produced (**Table 2.5**) and have been demonstrated to be selectively retained in different matrices (**Section 2.2.3**).

Since mammals and birds can generally metabolize many of these congeners, both transient and persistent metabolites may also contribute to biological actions from pulsatile PCBs in fish (See **Chapter 6**).

6 METABOLISM AND METABOLITES

Biotransformation of PCBs represents a small contribution to degradation and advection of total PCB in the accessible global burden. Nevertheless, relative proportions of congeners are changed. Furthermore, for the individual and biota in the immediate area or food web, this will have a substantial impact on body burdens and biological actions. Biotransformation also alters the toxicities of most chemicals, and PCBs are no exception. The biotransformation of several PCBs was determined in the early 1970s and general schemes and rules have been extensively reviewed and rediscovered. At the same time, some of the early observations were neglected in later, more focused, studies and this has led to some erroneous generalizations. Some of the earlier work must be, again, revisited.

The biotransformation of PCBs is obviously important in accumulation and persistence. Because of their lipophilicity, only small amounts of parent PCBs are excreted by vertebrates (Yoshimura et al., 1973; Yoshimura and Yamamoto, 1975; Haraguchi et al., 1998). Relatively larger amounts of even labile PCBs can be excreted unchanged by invertebrates with an open circulatory system (Saghir and Hansen, 1992; Saghir et al., 1993), but the majority of excretion depends on the introduction of more polar groups. Well-known exceptions are lipid-rich routes associated with reproduction such as eggs, breast milk and parturition.

6.1 Gone But Not Forgotten

Differential accumulation and retention of Aroclor peaks following experimental exposures (Borchard et al., 1974; 1975; Bush et al., 1974; 1975; Grant et al., 1971; Hansen et al., 1975; Platonow and Geisginger, 1973) correlated with congeners found to be dominant in environmentally derived residues (Jensen and Sundström, 1974; Kuroki and Masuda, 1977; Zell et al., 1978). These consistencies suggested which PCBs were labile or persistent. A host of studies identified mostly hydroxylated metabolites (see Sundström et al., 1976; **Section 6.3**). Studies with single PCB congeners further confirmed the structures amenable to biotransformation and excretion (Gage and Holm, 1976; Ghiassuden et al., 1977; Matthews and Anderson, 1975; Metcalf et al., 1975).

On the other hand, marked retention of non-parent radioactivity following administration of readily metabolized ^{14}C-PCBs (Brandt, 1977; Brandt and Bergman, 1987) suggested further roles for metabolism in PCB toxicities. Binding of oxidative products generated *in vitro* was repeatedly demonstrated (Wyndam et al., 1976; Morales and Matthews, 1980; Shimada, 1976), and interest was revived recently (Brouwer and van den Berg, 1986; Korach et al., 1989; Lans et al., 1994). Brandt, Bergman and colleagues also continued to pursue the biosynthesis and occurrence of sulfur-containing metabolites while parallel studies have implicated additional biological effects for these metabolites (Haraguchi et al., 1985; Nagayama et al., 1989; Kato et al., 1995).

The hydroxy metabolites shown to be persistent in rodent and human serum because of protein binding (Bergman et al., 1994; Shimada et al., 1985) may compete with endogenous compounds such as vitamins and hormones for transport binding sites (Brouwer and van den Berg, 1986; Lans et al., 1994). Some may have antiestrogenic and/or cytotoxic actions *in vitro* (Kramer et al., 1997; Moore et al., 1997), but are generally unavailable *in vivo* due to this binding (Hansen, 1998; Nagel et al., 1997). Body burdens of these persistent hydroxy PCBs have recently been shown to increase with age in parallel with CBs 138 and 153 (Sandau et al., 1998). The persistent sulfur-containing metabolites also constitute significant body burdens.

A major unanswered question is: What are the consequences of the more abundant metabolites from more lightly chlorinated or otherwise labile congeners? These metabolites are rapidly produced by enzymes induced in advance of phase 2 enzymes, bind *in situ*, and have totally different spectra of biological activities. Short exposures to minimal inducing doses may also cause prolonged exposures to metabolites. CBs 3, 15, 77 and 136 were found to bioconcentrate in the fetal intestine following a single modest oral dose to the dam on gestation day 18 (Lucier et al., 1978). The slowly metabolized CB 153 did not concentrate in the intestine. The putative glucuronides of hydroxy CBs were rapidly eliminated from the neonate intestine within 3-5 days postpartum, but were believed to be involved in enterohepatic recirculation of the fetus, cycling between the glucuronide and the free hydroxy CB (Lucier et al., 1978). Whole organism radiotracer studies, similar to those of Brandt (1977) and Klasson-Wehler et al. (1989;1996), have also confirmed that metabolite disposition is a meaningful question.

6.2 Mass Balance

Without mass balance or some accounting for unknowns, secondary metabolism can obscure meaningful conclusions even if meaningful questions are asked. The complex routes to methyl sulfonyl metabolites preclude their formation during *in vitro* microsomal incubations. On the other hand, several primary oxidative metabolites, especially of more lightly chlorinated congeners, undergo additional oxidations even *in vitro* (Jansson and Sundström, 1976; Kaminski et al., 1981; Amaro et al., 1996; McLean et al., 1996a; Oakley et al., 1996). Arene oxide formation also contributes dihydrodiol, oxepin, and keto-enol products (Kaminski et al., 1981; Koga et al., 1989; Sipes and Schnellman, 1987). Greb et al. (1975) identified 3 biphenylols and 2 diols of CB 18. Subsequent studies in a wide variety of species demonstrated that CB 18 yielded 7-9 polar metabolites compared to one or two for CBs 52 and 101 under identical conditions (Ghiassudin et al., 1976; Hansen et al., 1977; Hinz and Matsumura, 1977; Metcalf et al., 1975; Saghir and Hansen, 1992; Saghir et al., 1993).

More readily metabolized congeners also bind to macromolecules rapidly or are otherwise less extractable following oxidative metabolism *in vitro* or *in vivo* (Ghiasudden et al., 1976; Klasson Wehler et al., 1996; McClean et al., 1996b; Morales and Matthews, 1980; Saghir et al., 1993; 1994; Shimada and Sato, 1978; Shimada and Sawabi, 1984). Continued metabolism and reduced extractability help to explain why the expected and demonstrated comparative metabolism rates of, e.g., CBs 18 and 52 were reversed when only hydroxylated metabolites were quantitated following extended microsomal incubations (Borlakoglu and Wilkins, 1993). Therefore, for comparative (species and/or congener) metabolism studies, mass balance is critical and more reliable quantitative comparisons can be made based on congener disappearance.

Failure of mass balance due to unknown or unextracted metabolites is less critical and disposition is automatically included when relative congener "disappearance" is used. Changes in PCB congener profiles are potentially the most revealing for determining the most labile/persistent congeners, species/sex/age differences, and interactive effects under real-world conditions. The previously mentioned early studies demonstrating changes in congener profiles were accomplished with rather crude analytical separations. More refined specific congener analyses and a comprehensive understanding of profiles can be utilized to accurately describe species and congener differences in degradative metabolism (**Chapter 5**; Bedard and May, 1996; Brown, 1992; 1994; Bruhn et al., 1995).

6.3 Oxidative Metabolism

Hydroxylated products were the first to be identified (Block and Cornish, 1959; Hutzinger et al., 1972; Gardner et al., 1973; Moza et al., 1973; Yoshimura et al., 1973) and these products appeared to be more acutely toxic than the parent compounds (Yamamoto and Yoshimura, 1973). Arene oxides (Daly et al., 1972) were implicated as a major route of oxidative metabolism with the discovery of a dihydrodiol product of PCB 52 in rabbit urine (Gardner et al., 1973). Further evidence for the formation of a 3,4-oxide came shortly thereafter with deuterated 4-chlorobiphenyl (CB 3) (Safe et al., 1975). It was later shown that the 3-OH CB 52 could arise from direct hydroxylation without arene oxide formation (Billings and McMahon, 1978; Sipes and Schnellman, 1987). It was further shown that, if both *para* positions were occupied, NIH chlorine-shift products were likely (Schnellman et al., 1984; Sipes and Schnellman, 1987).

Further discussion of direct as well as arene oxide routes to hydroxylated products can be found in several reviews (ATSDR, 1997; Safe, 1989; Sipes and Schnellmann, 1987; Sundström et al., 1976).

Enhanced metabolism and elimination of more lightly chlorinated congeners has long been a general observation. The presence of vicinyl hydrogens (adjacent unchlorinated carbons) favors the degradative metabolism of higher chlorinated CBs (Berlin et al., 1975; Gage and Holm, 1976) and has been suggested as a means of "typing" PCB congeners (**Section 5.3.1**). Indeed, the consideration of vicinyl hydrogens helps to classify CBs according to rates of metabolism. Type-specific "metabolic slopes" further differentiate species according to metabolic capability toward PCBs as well as suggest states of CYP induction (e.g. Brown, 1992; Bruhn et al., 1996; Kannan et al., 1996; Looser and Ballschmiter, 1998). The application of metabolic slopes for various series of organisms and congeners is receiving increased attention.

Oxidation is not precluded, however, by the presence of *meta,para*-chlorines as evidenced by the rapid metabolism of CB 77 and the detection of products resulting from similar pathways (i.e., arene oxide and NIH shift). **Figure 6.1** illustrates some routes of metabolism, via arene oxide formation, independent of vicinyl hydrogens. The 4-OH-2,3,5,3'4'-pentachlorobiphenylol (bottom center) is the major hydroxyCB metabolite detected in rat and seal blood and is a major contributor to hydroxyCBs in human plasma (Bergman et al., 1994). In addition to CBs 105 and 118, this same metabolite could arise from the CB 156 3,4- or 4,5-arene oxide with the loss of a chlorine.

Although the HO-CB name of this metabolite could be "4-hydroxy-CB 109," CB 109 was produced in relatively minor proportions (**Appendix Tables 2** and **4**). Likewise the major parent for 4-hydroxy-CB 146 is probably CB 153. These two products, and especially 4-hydroxy CB 187 were the major retained metabolites in human whole blood from Inuits and a pooled sample from southern Quebec (Sandau et al., 1998).

Mono-*ortho*/di-*para* CBs such as 28, 66, 105 and 118 are generally highly persistent; however, they are more rapidly cleared from populations exposed to greater than ambient CAs (Brown, 1994). This is especially true for high TEQ exposures such as the Yusho and Yu-Cheng rice oils which contained elevated proportions of PCDFs (Brown et al., 1989; Chen et al., 1982; Tanabe et al., 1989). CYP1A oxidases, which catalyze the conversions represented in **Figure 6.1**, are more rapidly induced to higher levels than the other CYPs involved in PCB metabolism.

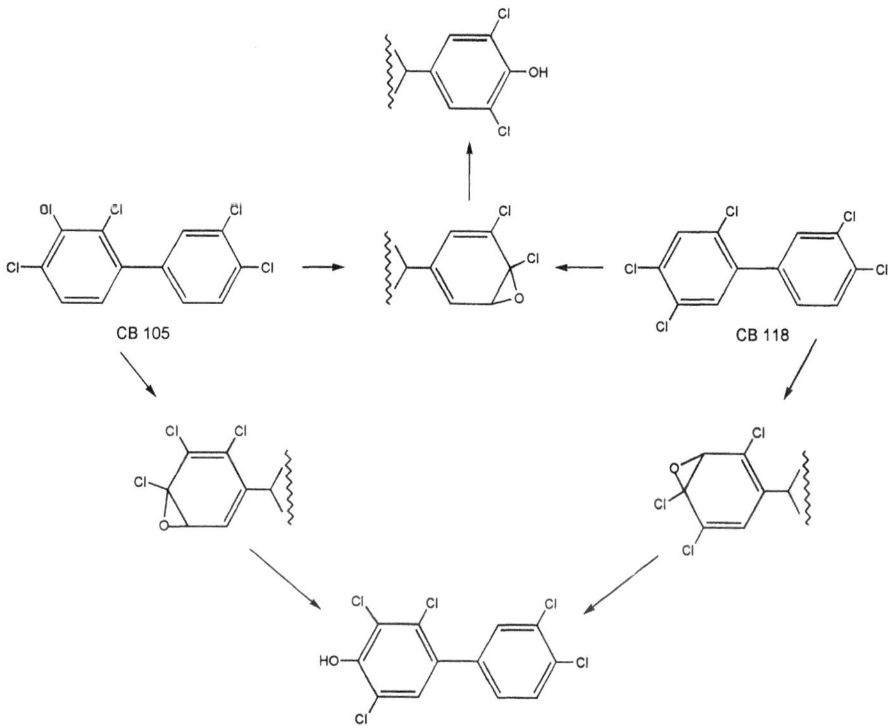

Figure 6.1. Arene oxide routes to hydroxyCBs resulting in chlorine shifts. The top and bottom (center) substitution patterns are commonly detected in mammalian tissues.

The 4'-OH-3',5'-diCl products (top center) are more common in human plasma than in seal or rat blood. The 4-OH-3,5,3',4'-tetraCB is a major metabolite of CB 77 in mice and rats (Klasson-Wehler et al., 1989; Koga et al., 1989). NIH shift products of the 2,4,5-ring have also been identified from CB 153 in the rat (Sipes and Schnellman, 1987). The mink metabolizes a significant portion of CB 101 via this pathway (Klasson-Wehler et al., 1996), but the rat (Haraguchi et al., 1997) or mouse (Klasson-Wehler et al., 1996) do not form detectable amounts. This might be due to more rapid metabolism of the lower chlorinated ring of CB 101 (**Figure 6.2**) by the rodents. Note that the dihydroxy CBs and the NIH shift product from the mink are not included in the abbreviated CB 101 metabolism scheme (**Figure 6.2**).

6.3.1 Routes of Oxidative Metabolism

Early metabolism studies demonstrated extensive hydroxylation of many PCBs. Some were concerned with less relevant 2,4,6-congeners such as CBs 100, 104 and 155 which helped to develop concepts (e.g., Goto et al., 1975; Lay et al., 1975; Sundström et al., 1976). More significant congeners were also studied; these included CB 4 (Moza et al., 1973; Goto et al., 1975), CB 15 (Safe et al., 1975), CB 18 (Greb et al, 1975; Ghiasudden et al., 1976; Hinz and Matsumura, 1977), CB 31 (Bakke et al., 1982; Felt et al., 1977; Moza et al., 1979), CB 66 (Yoshimura et al., 1973), CB 95 (Sundström and Jansson, 1976), and CB 101 (Chen et al., 1976; Gardner et al., 1976).

The very significant CB 153 was first reported to be totally refractory to oxidative metabolism (Hutzinger et al., 1972), but slow hydroxylation and dechlorination was later detected (Hutzinger et al., 1974; Jensen and Sundström, 1974). Failure to find oxidative metabolism of CB 153 by human liver (Schnellmann et al., 1983) was likewise later shown to be due to very slow production of metabolites (Ariyoshi et al., 1995). Extensive metabolism by the dog was shown to be due to a high level of CYP2B11 (See Sipes and Schnellmann, 1987; Ariyoshi et al., 1995).

CB 52 was extensively studied and provided an early model for metabolism pathways and metabolite toxicity (e.g. Gardner et al., 1973; Hutzinger et al., 1972; Norback et al., 1976; Preston et al., 1984), even in plants (Wilken et al., 1995). Further details can be found in nearly all previous reviews of PCB metabolism.

The coplanar CB 77 received early attention (Yoshimura and Yamamoto, 1973) and demonstrations of coplanar CB binding to the Ah receptor renewed interest in this congener. Although general PCB metabolite binding demonstrated by Shimada and Sato (1978) generated little immediate interest, rapid binding of CB 77 to microsomes and blood cells (Shimada and Sawabe, 1983; 1984; Shimada et al., 1985) did not go unnoticed. Darnerud et al. (1986) showed accumulation of phenolic and methyl sulfonyl metabolites in fetuses whose dams were administered the parent CB 77. Brouwer and van den Berg (1986) demonstrated binding of the major NIH Shift metabolite to transthyretin and this continues to be a pivotal consideration in PCB toxicities (e.g. Lans et al., 1993; 1994). Loss of Ah receptor activity by metabolism was demonstrated (Shimada, 1987; Yoshimura et al., 1987), but Ah receptor-independent effects were surfacing. The extensive metabolism of this congener was further studied and the additional metabolites elucidated advanced the understanding of potential routes of CYP 1A catalyzed oxidations (Klasson-Wehler et al., 1989a; 1989b; Koga et al., 1989).

CB 77 has been shown to be more rapidly excreted than CB 52 and, especially, than CB 47 (Clevenger et al., 1989; Gage and Holm, 1976; Shimada and Sawabe, 1984; Saghir et al., 1994). Nonetheless, Borlakaglu and Wilkens (1993) reported barely detectable metabolism even by induced rat and pigeon microsomes. Extensive secondary metabolism and tissue binding (Shimada and Sawabe, 1983) may have contributed to their misinterpretation as it did with CB 18 (see **Sections 6.2, 6.5**).

Addition of one more *meta* chlorine has a tremendous impact on metabolism. Only a single NIH Shift product of CB 126 has been identified (Koga et al., 1990) and essentially no transfer to the fetus occurs as with CB 77 (Klassen-Wehler et al., 1989b). Accelerated clearance of CB 126 is observed, however, in humans exposed to the high TEQ rice oil (Tanabe et al., 1989). CB 169 is generally considered unmetabolized. The mono-*ortho* pentaCBs and hexaCBs, however, are metabolized mainly via an arene oxide at the 3'4'- or 4',5'-position (**Figure 6.1**; Bergman et al., 1994; Haraguchi et al., 1998; Klasson-Wehler et al., 1989b; 1993). CB 118 can also form the NIH shift 4-OH product on the 2,4,5 ring as with CBs 101 and 153 (Haraguchi et al., 1998).

In spite of 3,4-arene oxides, the *para* chlorines generally decrease the rate of metabolism. The 2,5-ring has vicinyl hydrogens and is much more labile than the 2,4,5-ring. Two series of more completely studied congeners approach the 2,4,5,2',4',5' structure of the persistent CB 153 through different single chlorine substitution sequences and further illustrate the reduced metabolism of 4,4'-chlorination. CBs 15, 28, 47, 99, (153) are all 4,4' with increasing *ortho* then *meta* chlorination. All of these congeners are more refractory to metabolism than the corresponding homologs in the CB 4, 18, 52, 101, (153) series which are 2,2' with increasing *meta* then *para* chlorination (see **Figure 5.1**). Recall, though, that 3,4,3',4' (CB 77) is rapidly metabolized as a favored substrate for CYP 1A1 which it induces.

The oxidative metabolism of CB 101 illustrates the favored pathways for a single labile ring (**Figure 6.2**). This abbreviated scheme is supported by recently confirmed secondary metabolites to help establish routes.

The presence of two labile rings greatly expands the possibilities as can be seen by the subsequent summary of metabolic products from the very labile CB 95 (**Figure 6.3**). Pigeons, rats and mice hydroxylated every available position on CB 95 and some diol metabolites were also produced (Sundström and Jansson, 1976). The authors noted that pigeons tended to *para* hydroxylate the less chlorinated 2',5' ring of CB 95 *in vivo*, while rodents were more effective at *meta* hydroxylation of the more chlorinated 2,3,6-ring. A number of minor metabolites were also identified including three dechlorinated biphenylols and one dechlorinated biphenyl diol. Catechols and *O*-methylated products were destroyed by the sulfuric acid clean-up procedure, but catechols are quite probable through the pathways leading to dechlorinated biphenylols. The later demonstration of both 3- and 4-methyl sulfonyl metabolites (Bergman et al., 1992; Letcher et al., 1997) adds support to the

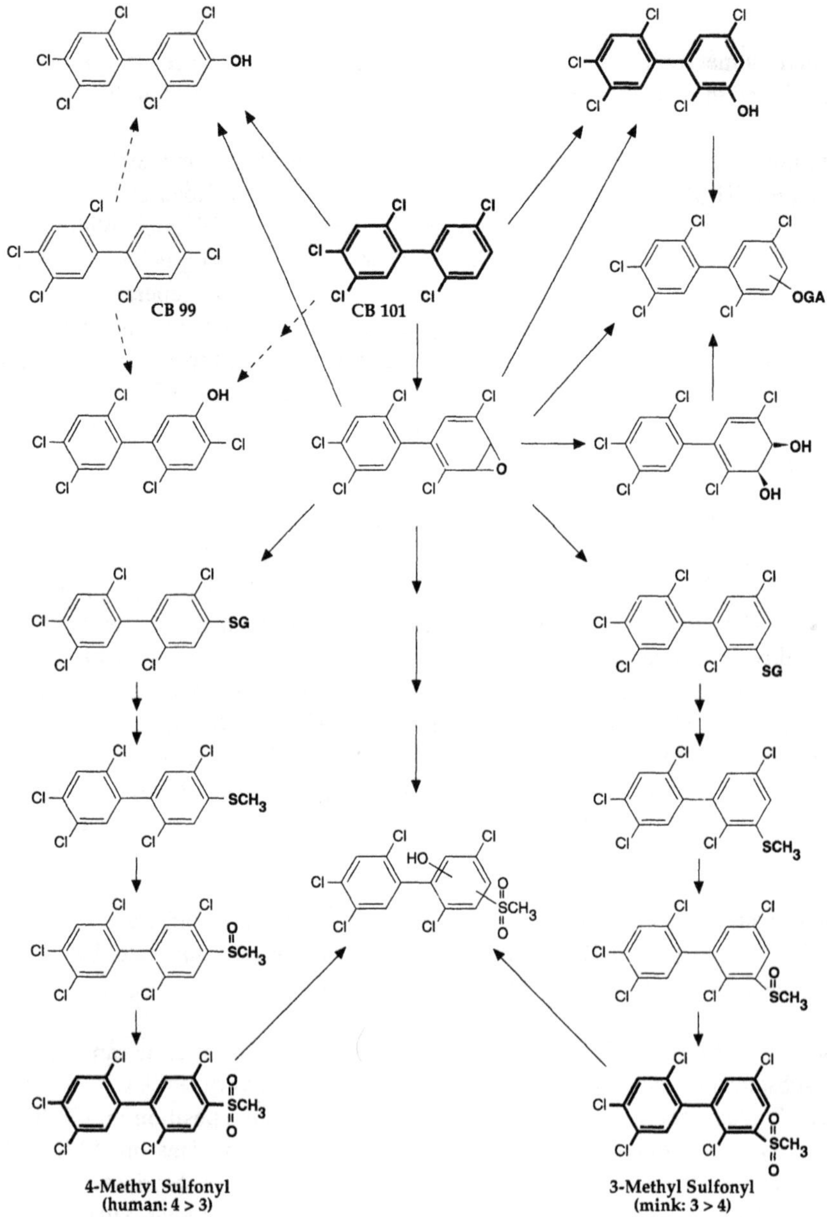

Figure 6.2. Representative metabolic pathways for CB 101 in various species. Bold structures are the major metabolites. Dihydroxy products and the 4-OH (NIH shift) product are not shown.

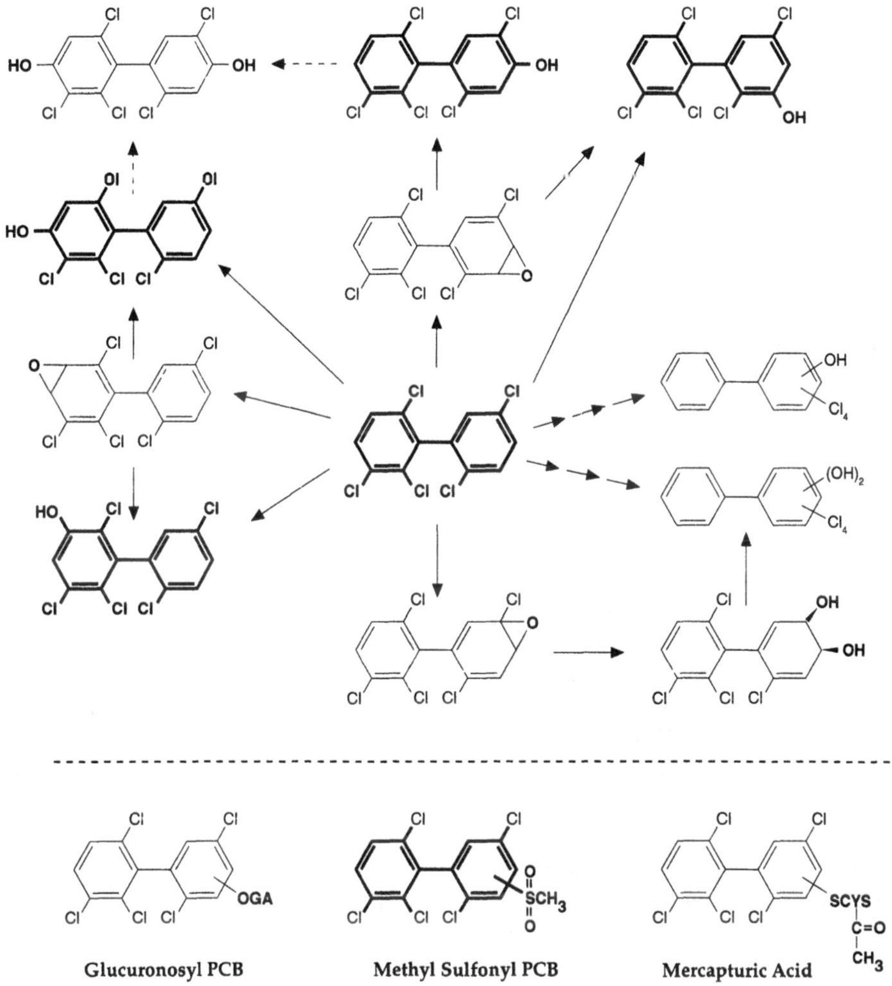

Figure 6.3. Metabolites and pathways for CB 95. As with **Figure 6.2,** major products are in bold. Because of the many products, a more generic presentation is necessary.

probable production of catechols by known pathways. Hydroxy-methyl sulfonyl metabolites are also probable *in vivo* (Haraguchi et al., 1987).

6.3.2 Enzymes of Oxidative Metabolism

Many xenobiotics induce their own metabolism as well as that of other exogenous and endogenous substrates. Chlorinated aromatics (CAs) are well-known to induce a broad profile of phase 1 and phase 2 enzymes and this has a significant influence on the disposition and toxicities of these compounds. Some biological actions are mediated by metabolites rather than parent CAs and induced metabolism of endogenous substrates causes other responses. The physiological consequences of arachidonic acid metabolism, for example, are many since the epoxides and hydroxyls participate in signal transduction. Changes in liver weight alone can disrupt thyroid hormone homeostasis (Byrne et al., 1987).

The primary enzymes involved in PCB metabolism are cytochromes P450 (CYPs). Over 100 distinct gene products have been identified in several different tissues. Several CYPs affected by PCBs and other CAs are presented in **Table 6.1**. PCB oxidation is catalyzed mainly by CYP 1A, 2B and 3A subfamilies and most of the work has focused on the hepatic enzymes. Other important sources include intestinal mucosa, nasal mucosa, lung and kidney. Most of these CYPs are membrane bound (endoplasmic reticulum isolated as microsomes) in close association with NADPH-cytochrome P450 reductase which is necessary for the oxidase activity. Thus, moderate size lipophilic chemicals have access to the enzyme, whereas large proteins and polar molecules are generally excluded as substrates.

The CYPs involved in steroid metabolism are found at high levels in adrenal cortex and gonads; some are microsomal, some are mitochondrial and some are in the cytosol. CYP 1B1 and CYP 3A4, for example, are inducible and catabolize steroids; the estradiol-4-hydroxylase in anterior pituitary (Bui and Weisz, 1989) may be the same as CYP 1B1 with the same activity inducible in MCF-7 cells (Spink et al., 1994). The major anabolic enzymes for steroids are less inducible than CYPs of the 1, 2 and 3 families; however, they bind xenobiotics such as PCBs and are subject to inhibition. The remainder of this section will deal mainly with the hepatic CYPs with broader substrate specificities which are most important in the oxidative metabolism of PCBs and other CAs. Nevertheless, induction of catabolic enzymes and inhibition of anabolic enzymes should be expected to disrupt profiles of steroid hormones.

Kaminski and coworkers (1981) convincingly established that dichlorobiphenyls which were *ortho*-substituted were metabolized more rapidly by PB-induced enzymes than *para*-chlorinated substrates were metabolized by BNF-induced P450s.

However, the CYP1A enzymes are induced to a greater extent by lower exposures than are the PB-induced enzymes so that elevated exposures to mixtures results in rapid metabolism of mono-*ortho* coplanars and CB 77 (Brown, 1994; **Section 6.3**).

The oxidation of *ortho*-poor PCBs by the highly inducible CYP 1A1 (Kaminski et al., 1981; Sipes and Schnellmann, 1987) accelerates the metabolism of otherwise refractory mono-*ortho* congeners such as CB 28, but not di-*ortho* congeners such as CB47; thus, following occupational exposure to PCB mixtures, CBs 15, 28 and 70 (and especially CB 77) may constitute a smaller proportion of the residue than in less exposed humans (Brown, 1994). A similar phenomenon of reduced CB 28 in higher dosed rats than in lower dosed rats has been observed (Ness et al., 1994). Fish have readily inducible CYP 1A but are frequently deficient in PB inducible forms; thus, they tend to accumulate more *ortho*-enriched residues (Hansen, 1998; **Chapter 5**). Fish (Stegeman et al., 1997) and bird (Gilday et al., 1998) CYPs are similar, but not necessarily identical, to analogous mammalian enzymes.

Many induction mechanisms are highly specific such as the gene expressions under AhR control; however, PB-type induction by diverse chemicals is still poorly understood (**Table 6.1**). It has been reported that PB and some PB-type PCBs induce several enzymes including CYPs 2A, 2B and 3A in concert (Lubet et al., 1992; Waxman and Azaroff, 1992; Larsen et al., 1994; Ganem and Jefcoate, 1998). CYPs 2B have been confirmed to oxidize *ortho* PCBs and CYPs 3A, with their very broad substrate specificities, have been shown to oxidize other chlorinated aromatics (den Besten et al., 1993; Ding et al., 1996). At this time, the author is unaware of reports directly confirming CYP3A metabolism of PCBs, but it should be strongly suspected.

Additional considerations of comparative metabolism were presented in **Sections 4.7, 5.3.1** and **5.4**.

Table 6.1. Common CYPs (P450s) induced by and/or metabolizing chlorinated aromatics (CAs) (Berenbaum et al., 1996; Ding et al., 1996; Feyereisen et al., 1989; Ganem and Jefcoate, 1998; Gilday et al., 1998; Huang and Gibson, 1991; Hung et al., 1996; Ioannides, 1996; Murayama et al., 1991; Spink et al., 1994; Spink et al., 1998; Stegeman et al., 1997).

TCDD[a)]	PB-Like[b)]	Common Assay[c)]	Degree Induced	Sample Substrate[c)]	Taxa
1A1		EROD	++++	Coplanar CAs, PAHs, steroids	Mammals
1A2		MROD	++	Amines; binds TCDD	Mammals
1B1			++	Steroids (E_2 4-OH)	Mammals
1A4			+	Coplanars	Chicks
1A5			+	Arachidonic acid	Chicks
1A-like		EROD	++		Fish
2A1	2A1[d)]	7α	+	7α	Mammals
	2A6	Coumarin	+	*ortho*CAs, Nitro cmpds	Human
2A8	2A8		+++	7α	Hamster
	2B1[d)]	PROD	+	16β	Mammals
	2B2[d)]	PROD,BROD	+	CAs	Mammals
	2B-like	PROD	+/-		Fish
	2B11	PROD	+	2,4,5 PCB ring, 16α	Canines
	3A4	BROD	+	Drugs, CAs, 6β	Human
	3A21[d)]	BROD	++	Drugs, CAs, 6β	Rat
	3A-Like		+/-		Fish
4A	4A		+,0,-	Arachidonic& Fatty Acids	Mammals
	6As		+	CAs, OPs	Diptera
6Bs			+?	Phyto-chemicals	Lepi-doptera

a) Generally via the AhR; CYP 2A1 and CYPs 4A are induced by coplanar compounds through mechanisms independent of the AhR. CYP 1B1 appears to be regulated by multiple mechanisms, including AhR.

b) Phenobarbital-like; not all induced by the complete spectrum of PB-like inducers.

[c] Commonly used assays and common substrates; steroids are indicated by the position oxidized (e.g. 7α, 15α).

[d] Subject to suppression by endocrine-dependent (T3, GH, etc.) mechanisms; levels elevated when suppression released. Very gender- and strain-dependent.

6.4 Phase 2 and Phase 3 Metabolism

In vivo, urinary metabolites of chlorinated aromatics are generally not extractable/detectable without acid or enzyme hydrolysis (Hinz and Matsumura, 1977; Klasson-Wehler et al., 1989a; Yoshimura and Yamamoto, 1973). This is due to the increased polarity and hydrophilicity of some phenolic and most conjugated products.

Unlike the urine, parent PCBs can be found in the feces and need not represent ingested compound that was not absorbed. Parent halogenated aromatics may be excreted through the intestinal mucosa, with its rich blood supply, into the intestinal lumen and appear in the feces even after non-oral routes (Yoshimura and Yamamoto, 1975; Sundlof et al., 1982). In addition, however, a complex profile of metabolites bound to various molecules can cycle into and out of the intestinal tract (see Klasson-Wehler et al., 1996). The origins of many of these complexes are phase 2 metabolism of phase 1 primary oxidative products. Both parent PCBs and metabolites have been found in high concentrations in salivary glands (Berlin et al., 1975; Klasson-Wehler et al., 1996) providing a secondary gi source for systemic-origin PCBs and metabolites.

Following phase 1 oxidations, the products may form biosynthetic conjugates with endogenous intermediates such as glucuronic acid or sulfate in what is referred to as phase 2 reactions. The endogenous intermediates require activation to high-energy forms (UDP-glucuronic acid; 3'-phosphoadenosine-5'-phosphosulfate or PAPS) and the appropriate enzyme. UDP-glucuronosyl transferases are generally within the endoplasmic reticulum in close association with CYP P450s while sulfotransferases are generally in the cytosol and, thus, are more likely to conjugate more polar metabolic products. Both are inducible. The glutathione transferases (GSTs) are mainly in the cytosol but some microsomal forms are present; a more reactive electrophile (such as an epoxide) is generally required since the glutathione-SH is less reactive than UDP-GA or PAPS. GSTs are also inducible.

The phase 2 reactions are very important in the toxicology of numerous xenobiotics but will be only briefly considered. Phase 2 products are generally more polar and less bioactive than phase 1 products. Most of the glutathione and cysteine conjugates undergo further metabolism to be excreted in the bile as mercapturic acids. β-lyases in the kidney and, especially, from the gastrointestinal flora can further cleave these conjugates to thiols which may undergo "phase 3" metabolism which is discussed in more detail below.

6.4.1 Polar Conjugates

Microsomal glucuronidation of oxidative products is well-established. It is also possible for some of the more polar metabolites to partition into the cytosol and be conjugated with sulfate. Detectable parent PCBs are not found in the urine and the polar conjugates are generally unextractable with the usual nonplanar solvents. Biliary conjugates may be degraded in the gastrointestinal tract and hydrolytic treatment is generally not necessary, but acidification can facilitate extraction and clean-up (Sundström and Jansson, 1975). Glucuronidation of hydroxy PCBs can also occur in the late fetus (Lucier et al., 1978; **Section 6.1**). Hydrolytic enzymes producing the aglycone can result in prolonged exposures at a vulnerable time due to the lack of an elimination route.

The phase 2 enzymes are generally induced at slower rates than CYPs 1A, 2B and 3A (Grote et al., 1977; Owens, 1985); therefore, within 2-5 days following exposures, oxidative products may be generated more rapidly than the phase 2 enzymes can conjugate them. This has been proposed as one factor contributing to nonlinear dose response relationships (Li et al., 1994) and should be investigated to account for different sets of responses following pulses of exposure verses equilibrium conditions.

For especially rapidly metabolized CBs and/or in highly induced tissues, the reactive oxidation products may bind nonspecifically to cellular constituents (**Section 6.2; Section 6.4.3**).

6.4.2 Cysteine/Mercapturic Acid Pathways--Methyl Sulfonyl Metabolites

In the 1970s, sulfur-containing metabolites of a number of xenobiotics were identified (see Bakke et al., 1982; Brandt and Bergman, 1987; Haraguchi et al., 1989; Mio and Sumino, 1985; Bergman et al., 1992; Norèn et al., 1996). Parent compounds included acetaminophen, caffeine and naphthalene (Kamei et al., 1976; Mio and Sumino, 1985); however, the most studied group was chlorinated aromatics. Propachlor and hexachlorobenzene were examples (Bakke et al., 1976; Jansson and Bergman, 1978), but DDE and PCBs dominated the early literature (Jensen and Jansson, 1976; Mio et al., 1976). Methyl sulfonyl PCBs continue to constitute a significant proportion of total PCB body burdens (Bergman et al., 1992; 1994; Letcher et al., 1995; 1997; Newsome and Davies, 1996; Norèn et al., 1996; Weistrand and Norèn, 1997).

Arene oxide activation of the halogenated ring is the major route to formation of cysteine and glutathione conjugates which then can follow sequential pathways to mercapturic acids, thiomethyls, methyl sulfoxides, and methyl sulfones (**Figure 6.2**). Because of the initial epoxide orientation, the 3 and 4 positions are the most common targets and NIH shift products are common. The mercapturic acids and cysteine conjugates are generally excreted in the bile and hydrolyzed to thiols by β-lyases in the intestine; the thiol is methylated and reabsorbed (Bakke et al., 1982; Mio and Sumino, 1985). For aromatic thiomethyls, the hepatic FAD monooxygenase mediates the conversion to a methyl sulfoxide while a P450 appears to add the second oxygen to form the methyl sulfone (Hajjar and Hodgson, 1980). PCB methyl sulfones are lipophilic and persistent, and are found at significant proportions of total PCB residues in several tissues and species (**Table 6.2**).

These metabolites are toxicokinetically as well as toxicologically significant: they inhibit lung P450s and AHH as well as induce PB-type CYPs in the liver (Kato et al., 1995; 1997; Lund et al., 1986; Nagayama et al., 1989). General effects on thyroid hormones and steroid receptors are becoming apparent. PCB $MeSO_2$s inhibit cell-cell communication with about the same potencies as parent CBs (Kato et al., 1998); however, the 3-$MeSO_2$s may be hundreds of times more potent in PB-type induction than the parent compounds (Kato et al., 1995; 1997).

6.4.3. Non-specific Binding of Phase 1 Products.

It is generally expected that rapidly metabolized PCBs will be bound to some extent to cellular macromolecules. The secondary oxidative products readily bind to nucleic acids as well as proteins (McLean et al., 1996a,b). Formation of fatty acyl esters and other lipid adducts is also possible. Mink excrete a larger proportion of lipid-bound CB 101 metabolites in the feces than do mice (Klasson-Wehler et al., 1996).

Table 6.2. Examples of methyl sulfonyl PCB metabolites identified experimentally or in environmental samples. The congener identified is the most likely parent compound.

B&Z No.	Ring[a]	Ring'	3-Me $S0_2$	3'-Me $S0_2$	4-Me $S0_2$	4'-Me $S0_2$	Sample Refs.[d]
6	2	3'			+		6
8	2	4'	++				6
8	6[b]	4'		+			6
31	25	4'			++++		1,4,5
49	24	2'5'	+		+++		4,8,10,11
52	25	2'5'	±		++++	(4,4')	4,5,8,10
64	236	4'	+		+++		7,10
70	25	3'4'	+		+++		4,8,10,11
77	34	3'4'	±		±		9
87	234	2'5'		+,++		+,+++	2,5,10,11
91	236	2'4'	+,++		+		10,11
95	236	2'5'	++				2,8
99	245	2'4'				101*[c]	
101	245	2'5'		+++		+,+++	2,4,8,10
110	236	3'4'	+		+++		2,10
132	236	2'3'4'	+,+++		+		10,11
141	2345	2'5'		+		++	10,11
149	236	2'4'5'	+,+++		+,+++		2,10,11
174	2345	2'3'6'		+		+	10,11

[a] Most-chlorinated ring as per IUPAC rules.

[b] Methyl sulfonyls are added, in some cases, in this table contrary to IUPAC rules (i.e. not on unprimed ring) to retain the numbering of chlorines of the parent CB. Since 2 and 6 positions are *ortho*, CB 8 is 2,4' OR 6,4'; 2,3,6 is correctly 2,5,6 if the first-named group (such as methyl sulfonyl) is present.

[c] Possible alternate source of product for indicated parent compound through NIH shift of the chlorinated arene oxide.

[d] The following references present the specific examples: (1) Bakke et al., 1982; (2) Bergman et al., 1992; (3) Bergman et al., 1994; (4) Brandt and Bergman, 1987; (5) Haraguchi et al., 1996; (6) Haraguchi et al., 1989; (7) Klasson-Wehler et al., 1987; (8) Letcher et al., 1995; (9) Darnerud et al., 1986; (10) Norén et al., 1996; (11) Weistrand and Norén, 1997.

6.5. Dose, Time, Induction and Biotransformation Rates

The primary variable determining rates of biotransformation for chlorinated aromatics is structure. In the real world, the co-variables of genetics, total dose, exposure rate, exposure duration, co-exposures and time also play major roles in the routes and rates of PCB metabolism. There are well-documented species (and strain) differences in biotransformation potential (**Chapter 5; Section 6.3**). The total dose may not be as critical as the rate, constancy, and duration of exposure. A minimum exposure level maintained for a specific time, including a single bolus, and a suitable lag for gene expression and protein synthesis are necessary for significant induction of both phase 1 and phase 2 enzymes; these factors are generally unique to each group of enzymes involved.

Occupational exposures to PCBs are generally sufficient to induce hepatic P450s adequately to enhance metabolism of the residues (Alvares et al., 1977; Brown, 1994). AhR-associated gene expressions, especially CYP1A1 and CYP1A2, are adequately sensitive and rapid to deplete some congeners from the residue profile so that accumulation factors may be lower than in the general population exposed to ambient PCBs (Brown, 1994). Persistent congeners such as CBs 28, 66, 105, 118 and 126 may be rapidly eliminated in highly induced humans (Chen, 1982; Tanabe et al., 1989) and in some fish; this results in unique profiles relative to those accumulated from ambient exposures. Induced P450s and the associated (expected) presentation of higher concentrations to the enzymes may also alter the routes of metabolism. Previous sections have demonstrated that several routes are possible and that secondary metabolism is the rule, rather than the exception, for PCBs.

Mass balance is extremely important in comparative metabolism studies. Early studies demonstrated multiple increasingly polar metabolites within the homologous series CB 18 >> CB 52 > CB 101 (Ghiasuddin et al., Hansen et al., 1977; Hinz and Matsumura, 1977; Metcalf et al., 1975). Recent studies by Borlakoglu and Wilkins (1993) confirmed low (relative to CB 52) production of monohydroxylated products of CB 18, but they failed to consider sequential metabolites in their extended incubation times (**Table 6.3**). No attempt was made to reconcile their low "metabolism" rates of CB 18 and CB 77 with the existing knowledge.

Table 6.3. Relative rates of CB disappearance compared to relative detection of hydroxylated products. In systems b and c there were 7 to 11 distinct spots for CB 18 and 1 major (+2 minor from sheep) spot for the other CBs.

		Relative Polar/Parent						
	Rat OH			Aquatic System[d]				
CB	Only[a]	Rat[b]	Sheep[c]	Algae	Snail	Culex	Fish
18	1.0	13.6	13.7	10.3	6.3	26.1	31.6
52	10.0	1.5	2.0	0.5	3.0	5.7	3.2
101	2.1	1.0	1.0	1.0	1.0	1.0	1.0

[a] Microsomal incubations measuring production of hydroxylated products (Borlakoglu and Wilkins, 1993).
[b] Microsomal incubations comparing parent to total polar products (Ghiasuddin et al., 1996).
[c] Microsomal incubations comparing parent to total polar products (Hansen et al., 1977).
[d] Laboratory model ecosystem organisms compared on the basis of biodegradability index (polar products/nonpolar products) (Metcalf et al., 1975). For CB 18 a major TLC spot even less polar than the parent was detected in the algae extract. Sanborn and Metcalf (personal communication) indicated that the mass fragments were consistent with a sulfur-containing metabolite.

To avoid errors in interpretation of comparative metabolism studies, substrate disappearance is the more critical measurement. For demonstrating routes of metabolism, mass balance is still important to determine the contribution of the routes identified to overall biotransformation of the substrate. Mizutani et al. (1978) also failed to find methyl sulfonyl metabolites of CB 18 following a single dose to mice; more slowly metabolized higher chlorinated congeners yielded these mercapturic acid pathway metabolites. This should not be interpreted as slow "metabolism." On the contrary, the CB 18 was probably too rapidly metabolized by secondary oxidative pathways to be available for the slower glutathione conjugation. Beluga whales from the polluted St. Lawrence estuary had lower ratios of methyl sulfonyl metabolites:total PCB than did lower residue whales from the less polluted Hudson Bay; however, much of this could have been due to exposure of Arctic whales to the different profile of *ortho*-enriched congeners as per **Chapter 4** (Letcher et al., 1997).

Concentrations also play a major role *in vitro* as well as *in vivo*. Net percent *in vitro* metabolism of CB 18 was decreased as the concentration increased from 1 nmole to 50 nmoles/5 ml (Hinz and Matsumura, 1977). Disappearance of nonpolar-extractable peaks from Aroclor 1242 also decreased as the substrate concentration was increased from 50 nM to 1 mM (Hansen, 1979; Hansen et al., 1977). Individual peaks were metabolized more slowly within the Aroclor than were the pure congeners (Hansen, 1987).

There are also both subtle and dramatic examples of substrate inhibition *in vitro* (Hansen, 1979; 1987; Kaminski et al., 1981; van Vliet, 1990). Hydroxylated products may also inhibit microsomal metabolism (Schmoldt et al., 1977). *In vivo* metabolism can also be inhibited by high doses of Aroclors (Beebe et al., 1992) and sulfur containing metabolites (Lund et al., 1986; Nagayama et al., 1989). Some of the apparent inhibition may be due to post-transcriptional suppression of enzyme levels in addition to or instead of direct inhibition (Hahn et al., 1993; Pang et al., 1998; White et al., 1997). Therefore, net metabolism of PCBs is more than just the product of a rate constant *x* concentration.

7 DISPOSITION AND TOXICOKINETICS

Different rates of metabolism are only one factor in the congener- and species-specific bioaccumulation of PCBs. Various transfer forces are also at work within the organism as within the global ecosystem. These processes can be described by toxicokinetic models.

Toxicokinetics concerns the time course of absorption, distribution, metabolism and excretion of a toxicant by an organism. Metabolism (more correctly, biotransformation) and excretion can be combined to describe "elimination" while distribution and elimination can be collectively referred to as "disposition". Toxicokinetics uses the same principles developed for pharmacokinetics, but applied to toxicants rather than drugs (Klaassen and Rozman, 1991; Shargel and Yu, 1993). As with global models, changes in concentration or amount with time can be reduced to mathematical relationships and predictive models can be constructed yielding a series of parameters. These parameters are useful in comparing toxicants, species, exposure routes, exposure rates and a number of other variables.

As with physical "constants" for PCBs, caution must be exercised when comparing toxicokinetic "constants" among different studies. It is far better to compare CBs and species with approximate ranges or rank orders than to derive some values from one model and other values from a different model. It is very tempting to use the very easily understood "half-lives" for comparison, but if different models have been used they are not comparable. For this reason it is necessary to review toxicokinetic models in a very basic manner while presenting actual values and comparisons.

7.1. Toxicokinetics

Traditional toxicokinetics involves multiple sampling of a meaningful reservoir of the toxicant that is related to the disposition of the toxicant in other reservoirs. The blood (especially plasma) is the most popular reservoir and it represents the "central" compartment. Excrement, tissue biopsy, expired air, or other compartments may be sampled depending on the properties of the organism and of the toxicant as well as the objectives. The concentration vs time is plotted semilogarithmically and rate constants are derived by differential calculus (**Figure 7.1**). Most of the equations have been solved, so that the transformations are

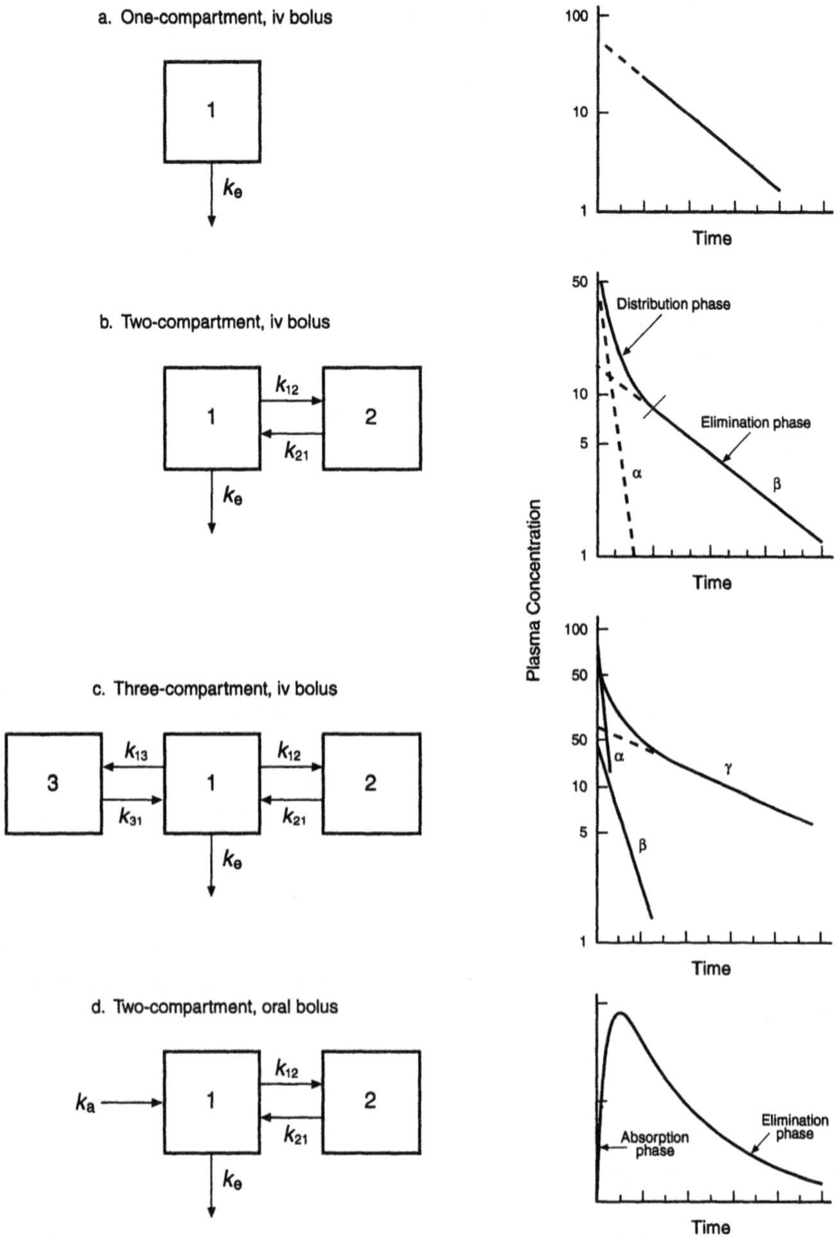

Figure 7.1. Examples of open-compartment toxicokinetic models for first-order processes and sample data plots. k_c is the elimination rate constant, k_{xx} are transfer constants between compartments and k_a is the absorption rate constant.

rather routine. Important relationships such as volumes of distribution (V_d), area under the concentration curve (AUC), and tissue or organism half-lives ($t_{1/2}$) can be determined by these methods. These values improve the understanding of the behavior of the toxicant in the organism.

Flow-based, perfusion-based or "physiological" based (PBPK) models may be derived from (or validated by) actual tissue concentrations, tissue volumes, and blood perfusion rates. PCB transport is rapid, but if transcellular transport is not very rapid relative to blood flow, a diffusion- or membrane-limited model is necessary. In PBPK models differential equations can be developed for all of the tissues, but the solutions become increasingly complex and difficult. Thus, it is common to combine rapidly equilibrated tissues (highly perfused) as one compartment and slowly equilibrated tissues as another. Tissues with unique distributions can be treated separately. Species extrapolations from PBPK models are more accurate than from the traditional toxicokinetic models. An example of a flow-limited model is presented in **Figure 7.2**.

Both compartmental and PBPK models are based on experimental data obtained under controlled conditions. The independent variable is time and the dependent variable is concentration(s). Covariables are usually rigidly controlled. In the real world, however, covariables may contain the most valuable information regarding sensitive populations, unique exposures, toxicant interactions, etc. Population-based (kinetic/statistical) models address these issues specifically and may combine separate sets of sparse individual data to create a rich data set (Sheiner, 1984; Steimer et al., 1994; Whittem, 1998). The "standard two-stage" (STS) method derives pharmacokinetic parameters and evaluates variability by statistical techniques. "Exploratory" data analysis is "essentially a mix of graphical and statistical techniques" (Steimer et al., 1994). Simulations can then be performed to approach the models providing the best explanations of the data.

7.2 Very basic toxicokinetic models

The basic units in toxicokinetics are concentration (mass/volume) and time. The experimental approach usually involves introducing a toxicant into an organism or system and measuring the concentration in some representative matrix at several time periods.

In certain instances such as iv infusion or saturated active transport, the amount of *A* transferred over time is constant and the relationship is linear ("zero order input"). Most toxicokinetic relationships of interest involve changes in concentration with time so that, if volume remains constant, the amount transferred is proportional to the amount remaining ("first order"). The relationship is log-linear and the actual first-order rate of loss can be arrived at by differential calculus. By taking samples at various times and plotting log concentration vs time, an equation for the straight line can be derived (after the appropriate transformations):

$$\log A = \frac{-kt}{2.3} + \log A_0$$

The 2.3 converts the log values to the natural log for the first-order rate constant, k, which is the slope of the line. A is the concentration at the time (t) measured and A_0 is the concentration at time = 0 (y intercept).

The first-order half-life is a constant determined from the slope:

$$t_{1/2} = \frac{0.693}{k}$$

Table 7.1 presents comparisons of k_e and $t_{1/2}$ reported for several PCB congeners. It is immediately apparent that direct comparisons, even for a single column of parameters, require some consideration of the sources of the values. It is generally known that the elimination of even the same CB from blood occurs at different rates depending on the time of sampling (i.e., there are multiple slopes to the elimination curve). The degree of resolution of these slopes strongly influences the reported kinetic parameters as is obvious from **Table 7.1**. The number of components determine the actual values of parameters as well as the kinetic model to be used.

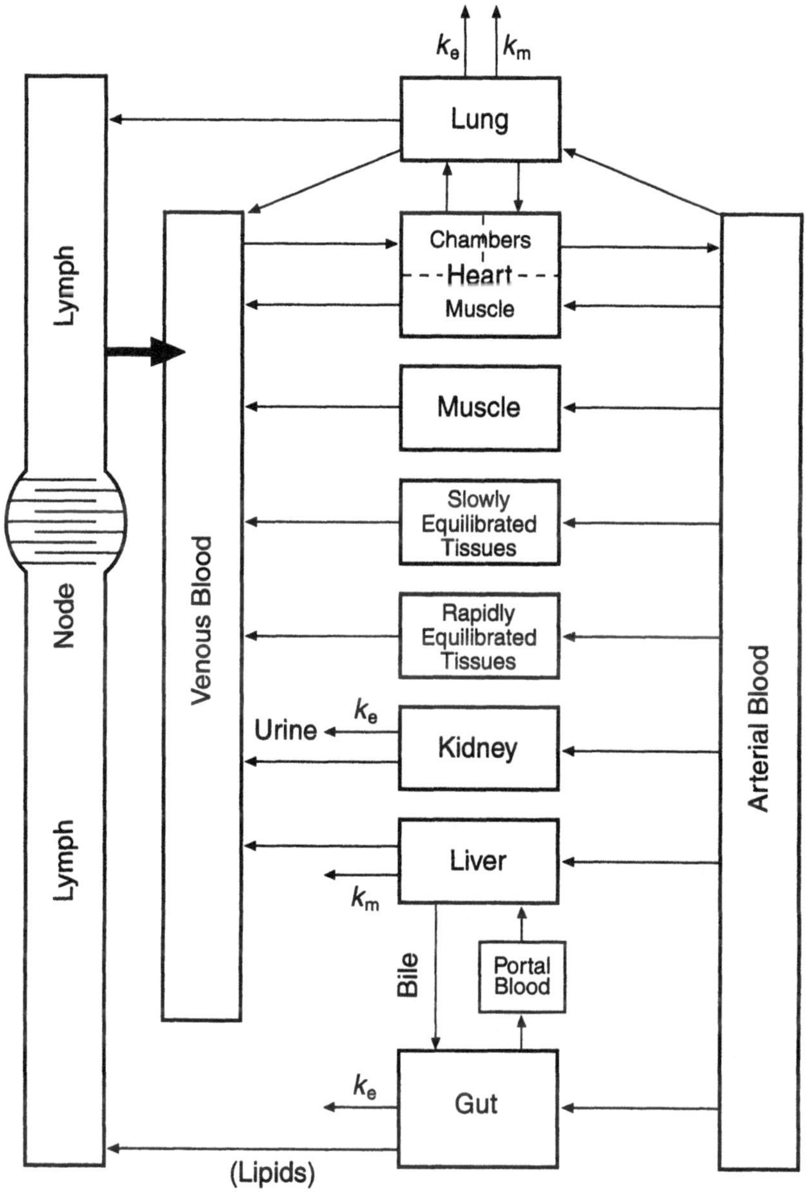

Figure 7.2. Example of a flow-limited or physiological-based pharmacokinetic (PBPK) model that might be applicable to PCBs. Important first-order rate constants are k_e (elimination) and k_m (metabolism). Flow rates, usually designated as Q, define the arrows from the central compartment rather than the transfer constants as in Figure 7.1. System inputs of toxicants (gut, venous blood, lung, skin, etc.) have not been included.

Table 7.1. Toxicokinetic parameters for various CBs determined from blood levels following iv administration. The number of resolvable components modelled is indicated by the numbers of entries.

B&Z No. & Ref.	k_{el} (day^{-1}) for Components			$t_{1/2}$ (day) for Components		
	1 (α)	2 (β)	3 (γ)	1 (α)	2 (β)	3 (γ)
3[a]	3.55	0.139		0.20	5.00	
15[a]	3.17	0.421		0.22	1.65	
28[f]	78.6	6.4	0.181	0.009	0.11	3.8
101[a]	31.7	0.600	0.065	0.022	1.15	10.70
136[b]	2.92	0.193		0.24	3.6	
136[c]	1.57	0.08		0.44	8.7[c]	
136[dd]	12.92	0.227		0.05	3.05	
136[dm]	4.26	0.152		0.16	4.57	
153[a]	4.42	0.612	0.012	0.16	1.13	56.62
153[b]	2.83	0.019		0.24	37.2	
153[c]	1.58	0.04		0.44	16.9	
153[ed]	2.67	0.045		0.26	15.0	
153[em]	2.30	0.015		0.30	46.0	

[a] Adult male Sprague-Dawley rat (250-300 g). CBs 0.6 mg/kg iv (Matthews and Anderson, 1975). Initial rapid (< 15 min) distribution from blood not resolved; 3 components resolved for CBs 101 and 153.

[b] Adult Sprague-Dawley rat (250-300 g). CBs 0.6 mg/kg iv (Matthews and Tuey, 1980). Distribution phase(s) not resolved; 2 components.

[c] Senescent (24-month) male Sprague-Dawley rat. CBs 0.6 mg/kg iv (Birnbaum, 1983). The $t_{1/2}$ for CB 136 (β) was not reported, but was calculated from data that were reported.

[d] Adult male beagles[dd] or cynomolgus monkeys[dm] given 0.6 mg/kg iv (Sipes et al., 1982a). Parameters for parent compound, metabolites reported separately.

[e] Adult male beagles[ed] or cynomolgus monkeys[em] given 0.6 mg/kg iv (Sipes et al., 1982b). Parameters for parent compound, metabolites reported separately.

[f] Adult male Sprague-Dawley rats (240-260 g) given 10 mg/kg iv (Moir et al., 1996). Early distribution phase resolved accounting for short α t½ (13 minutes).

7.2.1. One-compartment open model

If log plasma concentrations of a toxicant measured at several times following a bolus iv dose yield a straight line vs time, the equation and its solution are simple. The organism can be considered to be behaving as a single homogenous unit. Although this does not necessarily mean uniform tissue concentrations, changes in plasma concentrations are assumed to be proportional to changes in tissue concentrations. Because the toxicant is being eliminated, this is an "open" model. The elimination rate constant, k_{el}, represents the slope of the line (**Figure 7.1a**).

Longer-term PCB elimination may not be readily resolved into more than a one-compartment model if the data intervals are too great. It is also possible to miss very early and rapid distribution events (**Table 7.1**). It is known that a single compartment does not completely describe PCB disposition, but the larger data intervals "average" separate events. The simpler model provides a summary of the long-term disposition. This may be all that is necessary, or may even be preferred for certain considerations.

Excretion is usually most rapid when blood levels are highest and may be extremely slow once the initial PCB is eliminated from the central compartment by excretion and sequestration in fatty tissues. Models developed (and parameters determined) from non-invasive cumulative excretion are frequently limited to one or two compartments (**Table 7.2**).

7.2.2. Multi-compartment open models

If the plot of log concentration vs time yields a curved line, multiple compartments must be considered in the model. Various computer programs and even visual inspection can determine whether the data resolve 2,3, or more different slopes. The resolution capacity of the data is important. Even if logic supports additional compartments, data collected at too few time points restrict the model to the number of compartments (inflections) resolvable. The parameters in **Tables 7.1-7.3** were determined from 1-, 2-, or 3-compartment models.

Models of tissue disposition can be constructed from sequential samples. An absorption phase (into the tissue from the central compartment) can be defined so that a data plot similar to **Figure 7.1.d.** is obtained. Disposition from tissues such as liver, lung and kidney may include redistribution, metabolism and excretion components (**Figure 7.2**) and, thus, require biexponential or triexponential equations. For adipose tissue, a monoexponential equation (one compartment) should describe the redistribution from the fat once steady state is achieved. In some cases, 2 components for the redistribution of total radioactivity from fat may be resolved (**Table 7.3**). The first phase probably represents early more rapid removal of metabolites which accumulated because of higher concentrations in the central compartment during the tissue's absorption phase. In some cases such as with CB 153, k_{21} is so much slower than k_{12} that elimination cannot be accurately determined (**Table 7.3**).

Table 7.2. Elimination parameters for various CBs determined from excretion following iv administration.

B&Z No. & Ref.	k_{el} (day^{-1})			$t_{1/2}$ (day)		
	Feces		Urine	Feces α	Feces β or U + F	Urine
	α	β				
3[a]	1.058			0.654		
15[a]	0.749			0.925		
18[d]	0.120		0.324		1.56	
47[d]	0.018		0.006		28.01	
77[d]	0.076		0.014		7.74	
101[a]	0.425	0.079		1.63	8.79	
136[b]	0.93	0.125		0.75	5.54	
136[c]	0.533		0.674	1.30		1.03
153[a]	0.336	0.026		2.04	26.75	
153[b]	0.59	0.031		1.17	22.36	
153[c]	0.543	0.014	0.226	1.28	NR	3.07

[a] Adult male Sprague-Dawley rat (250-300 g). CBs 0.6 mg/kg iv (Matthews and Anderson, 1975).

[b] Adult Sprague-Dawley rat (250-300 g). CBs 0.6 mg/kg iv (Matthews and Tuey, 1980).

[c] Senescent (24-month) male Sprague-Dawley rat. CBs 0.6 mg/kg iv (Birnbaum, 1983).

[d] Weanling male Sprague-Dawley rat (67 $\pm$ 16 g). Dosed iv with 10.6, 5.3, or 2.2 mg/kg CBs 18, 47, and 77, respectively. Short-term (3-day) elimination (Saghir et al., 1999).

Table 7.3. Toxicokinetic parameters for elimination of various CBs from adipose tissue.

B&Z No. & Ref.	k_{el} (day^{-1})		$t_{1/2}$ (day)	
	α, β by Concentration	By Amount	α, β by Concentration	By Amount
3[a]	21.12		0.033	
15[a]	3.05, 0.329		0.23, 2.11	
28[f]	1.31, 0.10		0.5, 6.9	
52[d]	0.29		2.40	
52[e]	0.029	0.0064	24	108
77[d]	0.27		2.60	
101[a]	0.32, 0.043		2.14, 16.21	
118[e]	0.028	0.0026	25	268
Peak 1.27[e]	0.018	0.0040	38	170
136[b]	NR		5	
136[c]	0.27		2.5	
153[a,b,c]	NM[g]		NM	
153[e]	0.011	NM	63	NM
Peak 1.49[e]	0.017	0.0043	42	163
Aro 1254[e]	0.017	0.0024	42	284

[a] Adult male Sprague-Dawley rat (250-300 g) (Matthews and Anderson, 1975).
[b] Adult Sprague-Dawley rat (250-300 g). CBs 0.6 mg/kg iv (Matthews and Tuey, 1980).
[c] Senescent (24-month) male Sprague-Dawley rat (Birnbaum, 1983).
[d] Adult male mice (20-24 g) dosed po to pseudo-steady state with 8 x 100 mg/kg CB 52 and 10 x 8 mg/kg CB 77 Clevenger et al., 1989). Parent CB only.
[e] Weanling swine (19-21 kg) dosed po with 7 x 2.14 mg/kg/da (Hansen and Welborn, 1977). Peripheral fat biopsies assayed for parent CB only and results expressed both as concentration in fat and amount in total fat compartment (for dilution by growth). Peak 1.27 from Aroclor 1254 is mainly CBs 118 and 149 and peak 1.49 is mainly CBs 153, 105 and 132 (Table 3.1).
[f] Adult male Sprague-Dawley rat (240-260 g) (Moir et al., 1996).
[g] No meaningful value due to very slow elimination.

With multiple compartment models, k_{el} is not the simple rate constant as with the one-compartment model, even if elimination occurs only from the central compartment (**Figure 7.1b**). In the case of multiple-compartment models, the equation defines multiple straight lines and slopes include the transfer constants (k_{12}, k_{21}) as well as k_{el}. In a 2-compartment model, the initial slope is -α/2.3 and the terminal slope is -β/2.3. Mathematical analysis starts at the end, and the observed terminal slope is subtracted from the preceding total slope (observed values) to derive a net slope for the preceding component of the biexponential equation. Extrapolation of the terminal slope yields the y intercept, *B*, which is the plasma concentration intercept of the elimination phase; the difference between the extrapolated -β/2.3 (elimination phase) and the earlier data points yields the distribution phase. This can be extrapolated to obtain *A*, the y intercept of the distribution phase (**Figure 7.1b**). The sum of *A* and *B* is the theoretical plasma concentration at time 0.

For most CBs in **Table 7.1**, two compartments were resolved. When a third component is resolved, the terminal slope (γ) is more comparable to the terminal slope of the 2-compartment model (β). The k_{el} is still lower (half-life is longer) because the middle slope is resolved and no longer "averaged" with the terminal slope (CB 153 in **Table 7.1**).

The elimination rate constants determined for excretion (**Table 7.2**) are slower than elimination from blood (**Table 7.1**) and elimination from fat is even slower (**Table 7.3**). Fat is a "deep" compartment and the main or only way out for PCBs is to re-enter the central compartment (**Figures 7.1b** and **7.1c**). An increase in the relative contribution of the adipose compartment dilutes the PCBs and re-entry into the central compartment is even slower (Anderson et al., 1993; Birnbaum, 1983; Boon et al., 1984; Borchard et al., 1976; Hansen and Welborn, 1977; Mühlenback and Bickel, 1981). The long elimination times permit significant growth, including expansion of the fat compartment, so that concentration-based kinetics include both dilution and elimination. Estimation of the size of the fat compartment and determining body amounts rather than concentration in fat accounts for dilution and permits accurate calculation of elimination rates (**Table 7.3**). If dilution is factored out, actual elimination rates will be slower if the fat compartment is increasing in mass. The slower elimination of even CB 136 by older rats (**Table 7.3**) is due to lower perfusion rates and a larger adipose compartment which sequesters more PCB (Birnbaum, 1983).

Early toxicokinetic studies with PCB mixtures were restricted by short sampling protocols and packed column GC analysis. The advantage of these short-term studies was the demonstration of at least biphasic distribution of PCBs for the first 24 hours following iv injection (Borchard et al., 1974). The packed column peaks containing CBs 66 + 70 + 95, CBs 87 + 97, and especially CB 110 (**Table 3.1**) disappeared much more rapidly from sheep blood than the remaining peaks.

Disappearance from swine blood was nearly parallel for 8 peaks, but CB 110 was again more rapidly eliminated from the blood in 24 hours.

7.3. Models of oral exposure

First-order absorption rate constants (k_a) can be derived by operations similar to separating distribution and elimination phases for multi-compartment models (**Figure 7.1d**). The method of residuals or "feathering" involves subtracting extrapolated plasma concentrations from observed plasma concentrations to determine the residual concentration of interest.

Absorption of PCBs from the gi tract is quite rapid and elimination is slower. For some rapidly metabolized CBs, however, elimination may be more rapid than absorption so that the k_a may be mistakenly identified as the k_{el}. This "flip-flop" model is more often seen for rapidly eliminated drugs (Shargel and Yu, 1993) than for CAs.

PCB absorption may also be misinterpreted somewhat because of the assumption of intestinal transfer to the hepatic portal system. Absorption of digested fats occurs through lacteals rather than capillaries in the intestinal microvilli. Fats and fat soluble compounds enter the lymphatic circulation, joining the venous blood at the vena cava (**Figure 7.2**). It has long been known that fat-soluble drugs absorbed via this pathway avoid the first-pass effect of liver metabolism (Shargel and Yu, 1993).

As would be expected, absorption of lipophilic PCBs is intimately related to fatty acid absorption (Dulfer et al., 1998). Interception of lymph flow to the vena cava grossly decreases expected plasma PCB levels after feeding Aroclor 1242 to either dogs (Busbee et al., 1985) or sheep (Ziprin et al., 1980). PCB levels in the lymph were much higher than those in plasma, and interception by tissues other than liver before mixing with systemic blood may account for high levels observed in aorta, lungs and lymph nodes. Steady-state studies still rely on plasma concentrations and assume first-pass effects.

Some descriptions of the disposition of CBs following single oral doses were included in **Chapter 6**. Tissue distribution and elimination have also been described following single oral doses of CBs 153, 156 and 169 alone and in combination (de Jongh et al., 1993) and for CB 180 (Koss et al., 1993). Models and parameters describing the absorption of single oral doses of individual congeners, however, were not located.

Multiple oral doses have been used to determine times to steady state and maximum concentrations in mice (Clevenger et al., 1989; Morales et al., 1979). Steady states were achieved with CBs 3 and 15 within 2 days, but it was estimated that CB 101 would require at least 15 days and CB 153 at least 49 days to reach steady state levels from multiple oral doses (Morales et al., 1979). Clevenger et al. (1989) concluded that it would difficult, if not impossible, to reach steady state conditions with the rapidly metabolized CBs 52 and 77 due to enzyme induction, changing organ distributions and changing body sizes. Approximate steady states suitable for interpreting organ toxicities were achieved in 8 to 10 oral doses.

7.3.1. Absorption of PCB Mixtures

The absorption kinetics of single oral doses of individual PCB congeners do not appear to have been described. Single purified PCB congeners are useful for developing models and providing theoretical comparisons; however, the toxicokinetics of the components of actual mixtures are the ultimate concern. Since absorption from the gi tract is rapid and (most likely) avoids first-pass metabolism via lymphatic circulation, even poorly resolved packed column GC analyses can be used to describe PCB components in mixtures. This avoids the need to reconstruct the disposition of a theoretical mixture from the constants of only a few congeners.

Absorption rate constants for individual peaks (containing multiple CBs) in swine and sheep following single oral doses of Aroclor 1254 varied within a narrow range (Borchard et al., 1975). Apparent rate constants for accumulation in fat following feeding Aroclor 1254 at 200 ppm for 35 days were slower, as expected (Borchard et al., 1976). The two studies are compared in **Table 7.4**, but the feeding study provides only apparent k_a values useful in comparing peaks and species.

For the single dose study, the k_a for the individual peaks varied with species (Borchard et al., 1975). The swine values were more comparable to those estimated in humans following fish consumption. The monogastric swine absorbed PCBs more rapidly than the sheep, but there was a lag period for both species (possibly related to lymphatic absorption and delayed mixing with blood). Peaks 0.70 and 0.84 were most rapidly absorbed by swine. Peaks 0.70, 0.99 and 1.05 were most rapidly absorbed by sheep while peak 84 absorption was decidedly slower (**Table 7.4**) Absorption half-lives for total PCB were 1.13 hours for swine and 3.83 hours for sheep, with maximum blood levels occurring between 4-8 hours for swine and at about 12 hours for sheep.

Table 7.4. Comparison of absorption rate constants for Aroclor 1254 peaks following a single oral dose (blood) or apparent k_a for accumulation in fat during 5-week feeding.

Peak RRT[a]	Major Congeners[b] (B&Z No.)	k_a (h^{-1})			
		Swine		Sheep	
		Blood[c]	Fat[d]	Blood[c]	Fat[d]
0.70	66,70,74,91,95	1.11	0.0054	0.88	0.0035
0.84	84,99,101	0.68	0.0046	0.17	0.0031
0.99	*87*,97	0.58	0.0030	0.68	0.0031
1.05	110	0.55	ND[e]	0.84	ND
1.27	118,149	0.56	0.0045	0.17	0.0029
1.49	105,*132*,*141*,153	0.58	0.0050	0.19	0.0019
1.76	138,163	0.37	0.0038	0.19	0.0021
2.08	*128,187*	0.39	0.0043	0.29	0.0033
2.53	*174,177,183*	0.36	0.0042	0.17	0.0018
2.86	*156,180*	0.23	0.0049	0.16	0.0015
3.32	*170*	0.32	0.0039	0.17	0.0018
	Total PCB	0.61	0.0047	0.18	0.0027

[a] Retention time relative to *p,p'*-DDE = 1.00 (Table 3.1).
[b] As determined by Webb and McCall (1972). Italicized congeners are *probable* components (Sissons and Welti, 1971; Hutzinger et al., 1974; Brown, 1994; Frame et al., 1996a) (**Table 3.3**).
[c] From blood samples after single oral doses of 15 mg/kg (swine) or 30 mg/kg (sheep) Aroclor 1254 (Borchard et al., 1975).
[d] From backfat biopsies during feeding of 200 ppm Aroclor 1254 for 35 days (Borchard et al., 1976)
[e] Not detectable.

PCB absorption from the human gi tract is also rapid and nearly always greater than 90% (ATSDR, 1997; McLachlan, 1993). The influence of dietary intake was graphically demonstrated in two studies which monitored PCB levels in human blood following a fish meal. Total PCB in the blood of two volunteers increased to 150% previous levels within 3-5 hours following a fish meal, but declined to previous levels in less than 24 hours (Kuwabara et al., 1979). Humphrey (1987) found serum PCB levels increasing 2- to 5-fold, peaking at about 10 hours after a fish meal. Concentrations returned toward baseline within 24 hours, but slightly elevated levels were detectable for 2-3 days in the volunteer having eaten the more contaminated lake trout.

The blood PCB profiles following fish meals were enriched in later peaks and depleted in pre-CB 118 peaks (Humphrey, 1987; Kuwabara et al., 1979). The altered profiles may result from selective absorption, more rapid elimination, or both.

7.3.2. Elimination of PCB Body Burdens

Elimination rate constants for individual peaks from the blood and fat of swine and sheep following the 35-day feeding are presented in **Table 7.5**. The blood elimination was described by a two compartment model. The fat data was described by monoexponential equations for all peaks since a single deep compartment was sampled. The distribution from blood (α) was particularly rapid for peak 1.25 in swine and for peak 2.86 in sheep. Major peak 1.49 was very slowly eliminated from sheep blood; the main congener is the highly persistent CB 153. This low resolution peak is decidedly dominant in human breast milk as well (Krauthacker et al., 1986). The CB 66 + 70 + 95 peak (peak 0.70) was more rapidly eliminated from blood of both species following 35-day feeding than following iv administration (Borchard et al., 1974). The CB 110 peak was barely detectable in either species (**Table 7.5**). In the feeding study, the length of exposure provided for considerable induction of CYPs (Hansen et al., 1977; Wilson and Hansen, 1975) which could account for the low level of the labile CB 110 in fat after 35 days. Steady-state concentrations of some CBs may require 6-8 weeks of exposure, but the resulting profiles from the 35-day feeding study (Borchard et al., 1976) show species-specific peak patterns nearly identical to those seen after 90-100 day feeding of 20 ppm Aroclor 1254 (Hansen et al., 1977; **Table 3.2**).

In **Table 3.2**, swine and chick peak profiles were compared to those of humans following occupational exposure or following regular consumption of fish and seafood from the New Bedford, MA area. Differences among peaks are not as great as differences among individual congeners because most peaks contained both labile and persistent CBs. The rapidly metabolized 2,5 and 2,3,6 chlorination patterns (**Chapter 6**) predicted the more rapid elimination of CBs 52, 101 and 136 (**Tables 7.1, 7.2, 7.3**). The persistence of peak 0.48 in swine and chickens (**Table 3.2**) may be partly due to the presence of the persistent CB 47 as well as the labile CB 52. On the other hand, CB 52 as a single congener was much more slowly eliminated by swine than by rodents (**Table 7.3**).

Table 7.5. Elimination rate constants for individual peaks containing multiple CBs from swine and sheep following 35-day feeding of 200 ppm Aroclor 1254 (Borchard et al., 1976).

Peak[a]	Swine (days^{-1})			Sheep (days^{-1})		
	α	β	k_{el} fat	α	β	k_{el} fat
0.70	3.10	0.124	0.079	0.62	0.082	0.029
0.84	0.54	0.029	0.024	0.42	0.012	0.018
0.99	0.80	0.038	0.030	0.38	0.012	0.013
1.05	BD[b]	BD[b]	BD[b]	BD[b]	BD[b]	BD[b]
1.27	1.14	0.032	0.053	0.24	0.018	0.017
1.49	0.82	0.033	0.022	0.14	0.002	0.010
1.76	0.83	0.025	0.022	0.44	0.023	0.008
2.08	0.82	0.029	0.026	0.35	0.009	0.011
2.53	0.62	0.026	0.023	0.16	0.006	0.004
2.86	0.71	0.037	0.026	2.07	0.025	0.006
3.32	1.22	0.043	0.034	0.17	0.001	0.004
PCB	0.691	0.033	0.027	0.30	0.017	0.010

[a] Retention time relative to *p,p'*-DDE = 1.00; major congeners listed in **Table 7.4**.
[b] Below the detection limit.

Oral exposure of a human volunteer to a mixture of ^{12}C- and ^{13}C-PCBs resulted in relatively high one-day blood levels of CBs reported as 108/118, 138, 153 and 180 (Buhler et al., 1988). Elimination rates were influenced sporadically by putative changes in blood lipids. They were also affected by dietary intake, as evidenced when the natural isotope surpassed the synthetic mixture. By normalizing, the following elimination half-lives were calculated:

CB 118	100-300 days
CB 138	321 days
CB 153	338 days
CB 180	124 days

Most PCB residues result from irregular multiple pathway exposures. In addition, the long elimination times allow for fluctuations in diet, health and other factors influencing disposition. Nevertheless, declines in specific components of PCB mixtures after a major source is interrupted provide important inferences regarding the disposition of individual CBs in real-world situations.

Acute exposure to Clophen A30 during a fire in a pulp mill provided estimates of the relative clearance for 10 congeners in the 12 most exposed workers (Luotamo et al., 1991). PCB concentrations were elevated above those in comparable cohorts exposed only through environmental sources (**Table 7.6**), but were lower than those having been chronically exposed or exposed sporadically through multiple mishaps. Serum concentrations of CBs 18, 33, 66, and 101 declined rapidly to or below background values even though fat concentrations remained elevated (**Table 7.6**). CBs 28, 171, and 183 declined more slowly while CBs 153 and 156 remained essentially constant. CB 74 concentrations increased in both serum and fat, further supporting the suggestion of dechlorination of some higher homologues by humans (Brown, 1994; Hansen, 1998; Wolff et al., 1982).

Table 7.6. Serum and adipose residues of selected PCB congeners from ambient sources and after accidental exposure to Clophen A30 (Luotamo et al., 1991).

B&Z No.	Cl Pattern	Serum (μg/liter)			Adipose (μg/kg)		
			Accident[b]			Accident[b]	
		Env[a]	1 da	1 mo	Env[c]	1 da	1 mo
18	25-2	0.09	1.0	0.06	0.2	23.5	2.4
28	24-4	0.02	0.5	0.17	1.1	6.1	3.8
33	2-34	0.04	0.3	0.01	0.2	11.4	0.8
66	24-34	0.13	0.6	<0.01	0.6	19.0	2.9
74	245-4	<0.01	0.03	0.13	3.8	3.4	6.3
101	245-25	0.01	0.13	0.01	0.1	1.1	0.3
153	245-245	1.2	NM[d]	NM	132	142	149
156	2345-34	0.1	NM	NM	7.5	14.8	14.3
171	2346-234	0.02	NM	NM	2.4	19.8	9.7
183	2346-245	1.4	NM	NM	12.7	30.6	19.9

[a] Mean age 51.2 yr; n = 30; environmental exposures assumed.
[b] Pulp mill capacitor fire; n = 12; 1 day and 1 month after exposure.
[c] Autopsy specimens; Southern Finland; n = 29; mean age 34.1 yr.
[d] Not measured.

Wolff et al. (1992) compared the above data, several other reports, and determinations of low resolution GC peaks in capacitor workers before and 46 months after use of PCBs in the factory was discontinued. Estimated half-lives varied over a wide range (**Table 7.7**), indicating that numerous factors including continued exposure were involved during reduction of PCB body burdens.

Table 7.7. Half-lives of low resolution PCB peaks in humans as estimated from various literature reports (Brown et al., 1989; Wolff et al., 1992).

Peak[a]	Estimated half-life (months)					
	A	B	C	D	E[c]	F
0.37	60	5		36	17	1
0.47	72	3				
0.70	Infin[b]	Infin	13	96	38	varied[d]
0.78	36	4				0.5
0.84	72		156		40	0.5
1.25	120		10		70	
1.46	Infin		420	336	149	Infin
1.60			7			
1.74	204		312	192	(94)[e]	1.6
2.80	120		Infin	84		0.5

[a] Peak retention time relative to *p,p'*-DDE = 1.00. Comparable to designations in **Tables 3.3** and **7.4**.
[b] Elimination time too long to calculate; infinity relative to other peaks.
[c] Brown et al., 1989. Monthly values adjusted according to original manuscript rather than rounded years as per Wolff et al., 1992.
[d] Contained both CBs 66 (decreasing) and 74 (increasing). Data in **Table 7.6**.
[e] CB 138 $\approx$ 78 months and CB 163 > 240 months.

The half-lives for group B (pulse of exposure in children) and group F (pulse of accidental exposure; Luotamo et al., 1991) are considerably shorter than those in occupationally exposed groups after the major source was removed. The half-lives determined by Buhler et al. (1988) were also much shorter, indicating probable continued exposure of these cohorts to higher than ambient levels. Steady-state residues should be composed of mainly the more persistent congeners, but apparent half-lives of labile peaks were also quite long. Reservoirs within the environment of the workers (see **Chapter 4**) would still be present and continued exposure, even to lower levels, would still make the half-lives appear to be longer by these methods.

Group C values were determined from multiple samples from Yu-Cheng patients after the rice oil poisoning in Taiwan (Chen et al., 1982). The very rapid elimination of mono-*ortho* CB 105 (peak 1.6) and CB 118 (peak 1.25) was attributed to highly elevated CYP1A activity due to the presence of PCDFs in the rice oil (Brown et al., 1989; Chen et al., 1982). Similar accelerated clearance of mono-*ortho* CBs 28, 66, 105 and 118 was reported for a single Yusho patient (similar earlier PCB poisoning in poisoning in Japan) (Tanabe et al., 1989). In the Yusho patient, coplanar CB 77 was also rapidly eliminated and CB 126 was less persistent than would be anticipated in an uninduced human.

7.4. PBPK Model of early disposition

Matthews and colleagues studied the toxicokinetics of several ^{14}C-CBs in rats to develop relationships for degree and position of chlorination (Lutz et al., 1977; Matthews and Tuey, 1980). They developed a PBPK model for distribution and elimination of individual CBs applied to rodents, dogs and monkeys (Lutz and Dedrick, 1987; Matthews and Dedrick, 1984). The parameters and relationships presented in this model help to explain many of the distribution and redistribution phenomena observed in longer-term residue studies. The species differences are also useful, most being related more to metabolism than to other disposition factors.

Traditional toxicokinetic parameters from the initial studies have been presented in **Tables 7.1 - 7.3**. These data have been adapted to the PBPK model and CB 153 concentrations in various tissues of the rat are presented in **Figure 7.3**.

Efforts are underway to create PBPK models for all PCBs based on parameters such as tissue:blood partition coefficients (Parham et al., 1997) and rates of metabolism (Parham and Portier, 1998).

Identical tissue:blood partition coefficients were calculated for CBs 44, 49 and 52 (Parham et al., 1997) and these are reasonably close to those observed by Wolff et al. (1982). Somewhat greater deviation from observed values is seen for CBs 95

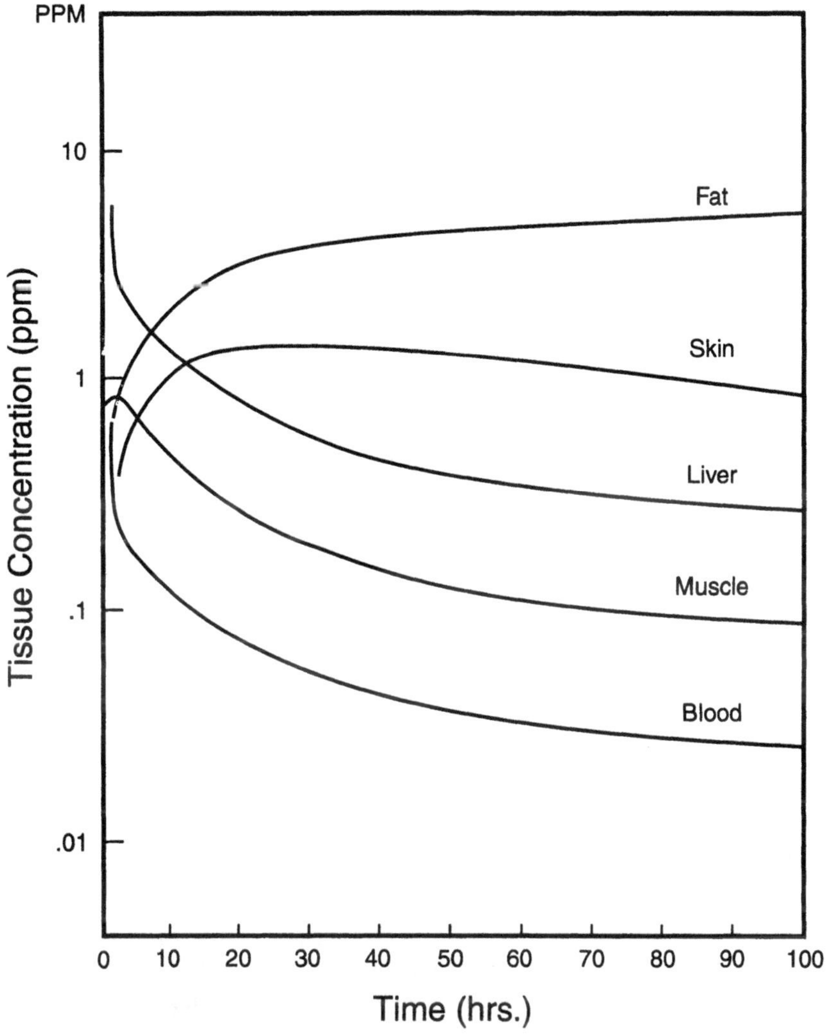

Figure 7.3. Short-term distribution of CB 153 in young adult male rats as predicted by a PBPK model and verified by analysis of total radioactivity following iv administration of 0.6 mg/kg ^{14}C-CB 153 (adapted from Lutz et al., 1977; Lutz and Dedrick, 1987; Matthews and Dedrick, 1984).

and 101, which also had identical calculated tissue:blood partition coefficients. The fact that CBs 47, 138, 146, 153 and 199 had identical calculated partition coefficients and similar reported values indicates that the model may accurately reflect certain observations. On the other hand, caution is warranted before global conclusions are based on necessarily simplified assumptions. Certainly, the composition of the lipids, themselves, influence the partitioning and the lipid composition is different among tissues, species, and developmental stages (Boon et al., 1984; Dulfer et al., 1998).

Recognizing that metabolic rates are key variables in a PBPK model for PCBs, attempts have been initiated to find the most relevant structural descriptors to determine comparative metabolism rates (Parham and Portier, 1998). Because of the profound effect of CYP induction on clearance rates of certain congeners (previous section), metabolism rates are being compared in both induced and uninduced animals. Unfortunately, limited *in vitro* data with poor mass balance distorts the predicted rates from those predicted from *in vivo* studies. Different sets of congeners are generally reported for the different types of studies so that some normalization is necessary. Systematic studies of substrate (i.e., congener) disappearance could provide better and more complete data than the selected-product-appearance studies on which they are relying.

Until this or a similar predictive model is developed and verified, the empirical incorporation of variables and subjective evaluation of comparative CB toxicokinetics must be used. In other words, risk assessment based on actual residues rather than piecemeal predictions is still necessary.

7.4.1. Longer-term disposition after acute doses

PBPK and traditional toxicokinetic models are generally derived from experiments that would be considered short-term when compared to the half-lives reported (e.g. **Table 7.7**). Dilution by growth and changing sizes/composition of the compartments are now almost universally considered. Conversely, shrinkage of the fat compartment may precipitate adverse responses due to mobilization of existing residues (e.g., Robens and Anthony, 1980; Villeneuve, 1975). Redistribution and other complicating considerations can be averaged by considering long-term trends, but there are certain developmental periods which are more vulnerable to the subtle effects of PCBs. Therefore, it is important to be at least vaguely aware of some perturbations in long-term trends which may cause sporadic exposures to higher-than-ambient or other-than-common PCBs. Two samples of long-term redistribution are presented.

Growing swine were administered 7 daily oral doses to a total of 15 mg/kg of Aroclor 1254, CB 52, CB 118 and CB 153; blood and backfat samples were assayed for up to 120 days (Hansen and Welborn, 1977). After a few weeks, calculated total body fat burdens of CB 153 and, to a lesser extent, CB 118 appeared to increase. This was probably due to redistribution from internal fat to the peripheral fat compartment which was sampled. Blood concentrations of CB 52 increased between 2 and 4 weeks and those of CB 118 increased between 4 and 8 weeks.

These deviations appeared similar to those observed by Buhler et al. (1988), but sporadic dietary intake was not a factor in the swine.

Infant mice were given single ip doses of 500 mg/kg Aroclor 1254; this dose had been found to result in promotion of initiated lung and liver tumors (Anderson et al., 1993). Growth dilution was also incorporated into this long-term study by using growth-phase specific body weight doubling times. Only 9 major congeners were quantified and the unique behavior of CB 138 suggested it was not resolved from CB 163. Two mono-*ortho* CBs (105 and 118) distributed most similarly to di-*ortho* CBs 99 and 128 and differently than mono-*ortho* CB 156; this may indicate coeluting pairs 105/132 and 118/149 were also not resolved.

Even accounting for these possible analytical explanations, rather complex relationships among carcass, liver and lung were seen during different phases of growth. Lung residues of CBs 128, 170 and 180 were low and declined to below detection limits within 1 week; lung residues of CBs 99, 105, 118, 153, and 156 declined in parallel with liver and carcass to 8 weeks, but did not change or actually increased during the next few weeks (Anderson et al., 1993). CB 138 (+ 163?) did not decline between 7 days and 84 days.

With only these two examples, it is apparent that single-time residue determinations and early toxicokinetic predictions are not yet adequate indications of specific exposures at any one point in time. Nevertheless, the effort required to derive more accurate estimates may only be justified for certain congeners above certain concentrations at specific times. Additional toxicity data are needed to determine the level of effort required.

The fact that PCB exposures are from multiple sources and may occur sporadically can be confirmed by examining the frequencies of detectable labile CBs in various large studies. The possible implications for adverse health effects are seldom addressed because of the overwhelming data and the tendency to average entire populations. Sporadic exposures of healthy adults to labile PCBs are probably of little medical importance; however, these same exposures during critical windows of development may have significant adverse consequences (Hansen, 1998). PCBs readily cross the placenta and are also bioconcentrated in the lipids of mother's milk.

Sporadic exposures may be most often associated with diet because of the species differences in toxicokinetics of PCBs. Some food web distinctions have been introduced in **Chapter 5** and are addressed in more detail in **Chapter 8**. Additional species-related toxicokinetic considerations are introduced as examples and to further document sporadic exposures to unique PCB profiles.

7.5 Toxicokinetics in Other Classes of Animals

Mammalian PCB residue profiles are frequently similar, but there are specific differences among the various taxa. The half-lives of CBs 136 and 153 in rats, dogs and monkeys (**Table 7.1**) illustrate some of these differences. Different orders, families and even strains of mammals have unique absorption and disposition characteristics (**Tables 3.2, 7.4, 7.5**) as well as metabolic capabilities (**Chapter 6**). As introduced in **Chapter 5**, there are major generalizable differences among the various vertebrate classes and other phyla of animals. These differences are important for both taxa-specific toxicities and food web transfers of PCBs and metabolites. A few basic examples are presented for perspective.

7.5.1. Fish

The profile of PCBs accumulated by fish has been repeatedly demonstrated to include significant proportions of congeners which are more labile in mammals (**Chapter 5; Appendix Table 10**). Net uptake involves mainly bioconcentration factors (BCFs) and food chain sources, designated by respiratory gain (R_g) and ingestion gain (I_g), respectively, in **Table 5.3** (Camphens and Mackay, 1997). Numerous studies have reported BCFs for individual PCBs from water (See Geyer et al., 1999).

There are definite species, trophic and habitat differences in PCB kinetics among fish so that overgeneralization may lead to erroneous conclusions. On a global scale, these differences probably average out (**Chapter 5**), but specific populations, ecosystems and individuals will be affected differently. The emerging concept of CB toxicokinetics in fish is that absorption is relatively structure-independent, coplanar congeners are slowly eliminated, but even modestly chlorinated *ortho*-rich CBs are mostly diluted by growth.

After exposing rainbow trout to ^{14}C-CB 52 for 36 hours, Guiney et al. (1977) demonstrated a rapid initial whole-body elimination ($t_{1/2}$ = 1.6 da) followed by an unresolved distribution phase and a very slow extended elimination period ($t_{1/2}$ = 970 da). After feeding channel catfish 20 ppm Aroclor 1242 for 84 days, there was modest elimination of total PCB from the edible carcass after 56 days on clean (extracted with fat-soluble vitamins added back) feed (Hansen et al., 1977). In both studies, redistribution of residues was more important than whole-body depletion (**Table 7.8**). Peak 0.48, containing CB 52 (**Table 3.3**) was also persistent in the catfish and total PCB declines resulted mainly from loss of peaks between 0.27 (CBs 17,18) and 0.37 (CBs 28,31) (Hansen et al., 1977); the congeners lost would have been mainly CBs 16, 32 and 54 (Brown, 1994).

Table 7.8. Retention of ^{14}C-CB 52 and total Aroclor 1242 in fish on clean diets. Values based on total amounts so growth dilution is not a factor.

Time (da)	Percent Initial (Whole) or Total Residue[a]							
	Whole		Muscle[b]		Fat		Offal[c]	Head
	CB	A	CB	A	CB	A	Aro	Aro
0	100	100	22	58	1.2	8	6.0	25.5
7	76	110	11	55	5.5	11	6.9	25.4
28	69	119	10	55	6.3	10	7.7	25.5
56	68	84	11	48	5.3	10	13.2	26.6

[a] In rainbow trout exposed to ^{14}C-CB 52 ("CB") (Guiney et al., 1976) or channel catfish fed 20 ppm Aroclor 1242 ("A" or "Aro") Hansen et al., 1977).
[b] Designated as muscle by Guiney et al. and composed of edible carcass in Hansen et al.
[c] Liver and kidney analyzed separately (1-2% total residue).

Given the slow elimination rates, these are relatively short-term studies. Even shorter elimination times were allowed for one kinetic study in zebrafish, but the k_as help to confirm CB non-specific absorption (Fox et al., 1994) as suggested by Niimi and Oliver (1983). Values from these two studies reporting CB accumulations from controlled exposures to specific congeners are presented in **Table 7.9** along with values from an 8-year depuration field study in eels.

Table 7.9. Toxicokinetic parameters for selected CBs in fish and eels.

B&Z No.	Zebrafish[a]			Trout[c]	Eel[d]	
	k_a (da^{-1})	k_e (da^{-1})	$t_{1/2}$[b] (da)	$t_{1/2}$ (da)	k_e (da^{-1})	$t_{1/2}$ (da)
3				40		
4				5		
14				15		
18	3760	0.2920	2.4	190		
31	3950	0.0867	8.0	196		

B&Z No.	Zebrafish[a]			Trout[c]	Eel[d]	
	k_a (da^{-1})	k_e (da^{-1})	$t_{1/2}$[b] (da)	$t_{1/2}$ (da)	k_e (da^{-1})	$t_{1/2}$ (da)
40				107		
44	3200	0.0461	15.0		0.0017	400
49	3200	0.0458	15.1		0.0018	380
52	3230	0.0387	17.9	500	0.0005	1370
70	3340	0.0280	24.8			
77	5160	0.0224	30.9	44		
87	3360	0.0140	49.5	155	0.0006	1160
97	3400	0.0126	55.0		0.0009	760
101	3850	0.0130	53.3	>1000	0.0005	1450
126	6490	0.0099	70.0			
128	4970	0.00843	82.2	>1000	NA el[e]	
136	4300	0.0161	43.0			
138	4770	0.00624	111.1		NA el	
141	4910	0.00764	90.7	850	NA el	
149					0.0002	3550
151	4220	0.0121	57.3		0.0007	950
153	4660	0.0104	66.6	>1000	NA el	
169	7200	0.00766	90.5			
180				>1000	NA el	
183	6000	0.0876	7.9			
187					NA el	
194	5640	0.00865	80.1			
209	3640	0.0132	52.5			

[a] Flow-through mixture of 28 congeners; 30-day exposure followed by 30-day elimination (Fox et al., 1994).
[b] Not reported, but calculated as $0.693/k_e$; the k_e was determined from the steepest (α) phase of the elimination curve.
[c] Rainbow trout, single oral dose, whole fish basis (Niimi and Oliver, 1983).
[d] Yellow eel from contaminated River Rhine marked and transferred to clean lake; 8 year observation (de Boer et al., 1994).
[e] No apparent elimination, when corrected for growth, over the terminal 7+ years of monitoring.

The profiles of PCBs in fish (**Appendix Tables 10, 11**) reflect the kinetic parameters in the trout and eel. In general, the *ortho*-rich homologs are much more persistent than the coplanar homologs in fish (compare CB 77 to other tetraCBs and even triCBs in **Table 7.9**). Coristine et al. (1996) also found a slower K_e for *ortho*-CBs, especially CB 101. When caged trout were suspended in the River Po (Italy) and fed a partially purified diet daily, predicted steady state concentrations were more consistent with reported profiles (Vigano' et al., 1994). CB 110 (which was not considered in the previous studies) reached steady state levels near those of CBs 118 and 153. The major congeners accumulated from contaminated river water (Vigano' et al., 1994) were similar to those generally accumulated in fish oils and retained by marine mammals (Boon et al., 1994; Muir et al., 1988; Schantz et al., 1993). These include CBs 101, 110, 118, 138, 149, and 153. CB 110, however, is not consistently retained in the marine mammals, similar to its low persistence in terrestrial mammals (**Tables 3.2, 7.4, 7.5**). CBs 52, 84, 95, 101, and 149 are accumulated to greater degrees in fish and persist to a greater degree in marine mammals. Although CB 105 appears to make a strong contribution in the fish tables, it is likely that over half the peak is CB 132 as in seals (see **Section 8.2**).

7.5.2. Birds

Some toxicokinetic parameters for birds have been estimated, but no formal toxicokinetic studies were located. Accumulation of PCBs in birds is mainly from the food chain. Net PCB profiles, therefore, are related to those of dietary sources. Although no absolute kinetic parameters were located, presentation of selected residue profiles associated with different food sources may be useful (**Table 7.10**).

PCB elimination from birds is similar to SARs for mammals. CYPs and CYP induction are unique to birds, but 2,3,6- and 2,5-chlorination patterns are still generally the least persistent of the higher homolog groups. At 20 ppm dietary exposures, laying hens accumulated 100X greater residues of CB 118 in the liver than those of CB 84, in spite of unspecified induction of P-450 activity by CB 118 and suppression of the same activity by CB 84 (Hansen, 1979). Cormorants did not accumulate CBs 52, 101, 110, or 149 in spite of elevated concentrations in their fish diet (Falandysz et al., 1998).

CB 101 accumulated in nestling tree swallows above model-predicted values (Nichols et al., 1995), and this may have reflected relative concentrations in the insects consumed. In the area of the Sangamo Landfill (**Table 4.5**), starling chick residues were 12-13% each of CBs 101 and 118; in starling adults, CB 101 was nearly 2-fold higher than CB 118 (Arenal, 1997). Shore birds from the Mediterranean generally had only 20 to 70% as much CB 101 as CB 118 (**Table 7.10**).

Table 7.10. Representative homolog composition (percent quantified congeners) of PCB residues in the eggs of tree swallows and birds from the Mediterranean.

	Tree Swallow[a]	Avocet[b]	Little Tern[b]	Herring Gull[b]	Audouin Gull[b]
(Diet)	(Insects)	(Inverts)	(Fish)	(Mixed)	(Fish)
TetraCl	15.6	13.0	6.2	1.4	0.6
PentaCl	20.2	12.6	16.8	6.3	8.2
CB 101	2.9	2.0	4.4	2.3	2.5
CB 118	2.7	8.4	7.5	3.2	4.9
HexaCl	26.0	28.5	38.9	38.2	39.9
CB 153		11.1	18.8	16.0	21.2
HeptaCl	15.8	25.7	30.9	38.2	37.1
CB 180	4.7	5.6	12.4	14.3	19.3
OctaCl	4.7	19.6	6.7	15.1	13.3
TOTAL (ppm)[a,b]	1.4	0.5	14.5	18.3	34.1

[a] Adapted from Nichols et al., 1995. Total PCB in μg/g fresh weight.
[b] Shore birds adapted from Focardi et al., 1988. Total PCB in μg/g dry weight.

Some PCB congener profiles from birds deviate markedly in highly specific manners, possibly due to analyte selection (**Chapter 3**) but probably more related to peculiarities of exposures. Insectivores such as starlings reflect PCB residues of the near-nest soils (Arenal and Halbrook, 1997) and the association can be traced to insects in the stomach contents (Arena et al., 1999). The limited tree swallow data reported by Nichols et al. (1995) were pooled and more analytes were reported by Froese et al. (1998). CBs 17 and 136 were present at high concentrations but low frequencies in the invertebrates sampled and this was presumably site-specific. In calculating the percent composition of the residues for swallows and insects (**Table 7.11**), I used the sum of reported ng/g lipid rather than the reported total PCB; CBs 17 and 136 were not included in this total because some manipulation for low frequency occurrence was obviously introduced by the authors. The remainder of the percent composition data, however, are uniquely distinct from opportunistic-feeding herring gulls and invertebrate-eating stilts (**Table 7.11**).

Table 7.11. Percent congener compositions of eggs as reported (Focardi et al., 1988)[a] or of eggs, nestlings, and insects derived from pooled data expressed as ng/g lipid (Froese et al., 1998)[b].

B & Z No.	Eggs			Swallow[b] Nestling	Insects[b]
	Stilt[a]	Gull[a]	Swallow[b]		
16 + 32			0.4	0.9	1.3
17[c]					0.6% sPCB
28 + 31			12.5	12.1	6.4
42			4.4	6.2	21.2
44	1.5				
49	1.3				
52	4.4				
60	2.6	0.1			
61 (?)[d]	3.3	0.2			
66 + 95	8	1.1			
99	1.7	0.7			
101	4.4	2.2	NR[e]	NR	NR
110+77	3.1	0.3	9.8	11.9	14.8
118	8.9	3.1	12.0	14.2	11.0
128	0.8	2.5			
136[c]			17% sPCB	6% sPCB	23% sPCB
138	9.7	14	17.8	18.8	13.5
146	3.5	3.8			
149+123			7.3	7.5	9.4
151			2.0	1.9	2.7
153+132	12.6	16	17.1	16.7	11.9
156	1.7	1.9			
170	4.8	11.4			
172	1.3	2			
174			2.4	2.7	2.1
177	0.8	1			
180	8	14.2	0.6	0.5	0.3
182+187	5.8	5.1	13.7	6.5	5.4

B & Z No.	Eggs			Swallow[b] Nestling	Insects[b]
	Stilt[a]	Gull[a]	Swallow[b]		
183	1.7	4.5			
194	3.5	5.7			
196	2.6	5.4			
201	2.8	4			
Total[c] (ppb)	480	18310	1500	1600	1900

[a] Major congeners reported from eggs of Mediterranean shore birds (Focardi et al., 1988). Residues as dry weight of eggs.
[b] Congeners reported as ng/g lipid (Froese et al., 1998, related to Nichols et al., 1991). Because of large variations due to collections from distinct sites, average percent compositions were calculated from totals of mean congener concentrations rather than the reported total PCB value.
[c] CB 17 and CB 136 were present only in some of the samples and values were not included in total quantified congeners (see footnote[b]). Their contribution was determined as a percent of the reported total PCB (sPCB).
[d]CB 61 (2345) is quite rare and may reflect lab contamination or misidentified congener. It may also be unique to Italian PCBs (Fenclor) since it was also reported by Zuccato et al., 1999).
[e]Not reported; CB 101 concentrations were similar to those of CB 118 in the original data set (**Table 7.10**).

Egg yolk is an important route of excretion which is more constant in domestic layers than in wild birds; thus, seasonal layers would accumulate relatively higher body burdens at different times. In a given clutch of great tit eggs, the first 6 had PCB residues 2-3 times higher than the last 6 (Winter and Streit, 1992). The importance to food chains is illustrated by PCB exposure of Native Americans living in a noncontaminated area. The source was discovered to be annual consumption of the eggs of loons migrating from the Chesapeake Bay (Clarkson, 1995). The PCB content of market eggs is generally low and the profile in domestic eggs would be expected to be similar to that of poultry and dairy products. Of course, isolated incidents may result in sporadic exposures through this route.

7.5.3. Insects

The total biomass of arthropods is highly significant in aquatic, marine and terrestrial environments and should contribute substantially to PCB mobilization, metabolism and bioaccumulation by predators. PCB profiles in aquatic and marine invertebrates have been briefly mentioned in **Chapter 5**. Several authors have described uptake of specific PCBs by aquatic insects and the subsequent contributions to food chains. Insects also retain polar metabolites of labile CBs and contribute these, as well, to food chains. For example, radioactivity retained by mosquito fish (*Gambusia affinis*) partially reflect the metabolite profiles found in mosquito larvae (*Culex pipiens*) in a model ecosystem containing ^{14}-C congeners (Metcalf et al., 1975; **Table 6.3**). Salt-marsh caterpillars (*Estigmene acrea*) as well as houseflies retain as well as excrete significant amounts of polar PCB metabolites (Metcalf et al., 1975; Saghir et al., 1993; 1994).

Terrestrial insects have not generally been considered in great detail, but toxicokinetic studies in house flies provide sample information. Insects accumulate toxicants from diet, contact and aerosol exposures as evidenced by achieving satisfactory control of target species by these routes. Crickets suspended over contaminated soil accumulated greater than 1 ppm PCB in 3 days (Paine et al., 1993). Various caterpillars biomagnified PCBs about 10x from oak leaves (Winter and Streit, 1992). The most reliable administration is by topical application. Even though this may not contribute the greatest proportion of environmentally-obtained body burdens in flies, topical application provides insights into the relative toxicokinetics of individual CBs. **Figure 7.4** presents a simplified model of PCB absorption, disposition and elimination by house flies.

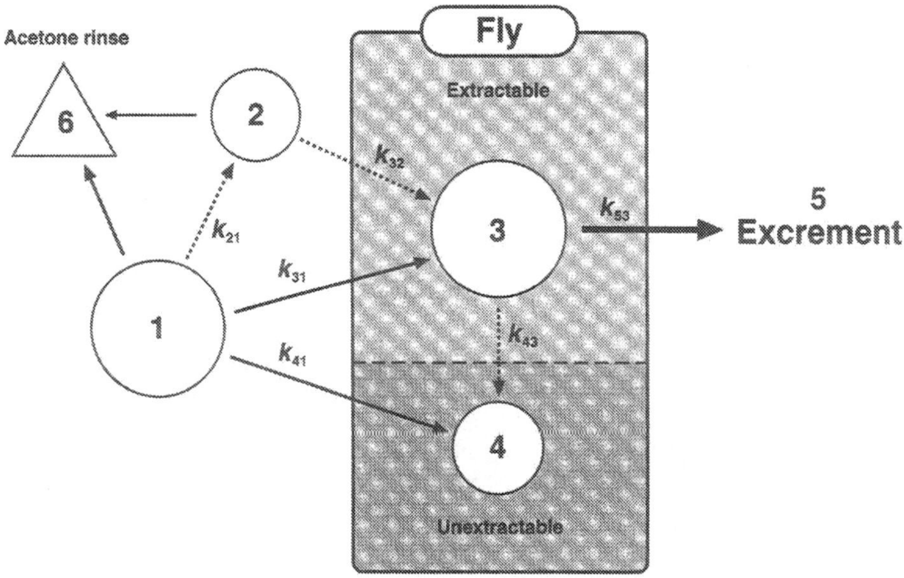

Figure 7.4. Model of PCB absorption and disposition following topical administration to the house fly, *Musca domestica* (Saghir et al., 1993, 1994).

The complex insect integument is represented by compartments 1 and 2, while unabsorbed PCB recovered in a brief acetone rinse at selected times is represented by component 6 (**Figure 7.4**). Absorption of CBs is apparently accomplished via two independent routes which can be modelled as slow (k_{21} + k_{32}) and fast (k_{31}) processes. Recent studies have demonstrated lateral diffusion within the cuticle and selective disposition within the fly (Saghir and Hansen, 1999); however, the open circulatory system of arthropods made it difficult to resolve more than the single internal compartment 3. Excretion from this compartment is represented by the rate constant (k_{53}), and total elimination includes a time- and dose-dependent accumulation of unextractable radioactivity in compartment 4 represented by rate constant k_{43}. Sample values for these parameters after topical application of 10 μg ^{14}C-CBs 18, 47 and 77 are presented in **Table 7.12**.

Table 7.12. Toxicokinetic parameters in adult female house flies following topical application of 10 μg/fly ^{14}C-CBs 18, 47 and 77 according to the model in **Figure 7.4** (Saghir et al., 1993; 1994).

B&Z No.	Absorption			Elimination		
	k_{21} (hr^{-1})	k_{32} (hr^{-1})	k_{31} (hr^{-1})	k_{43} (hr^{-1})	k_{53} (hr^{-1})	$t_{1/2}$[a] (hr)
18	0.0022	0.0005	0.168	0.00653	0.09573	5.2
47	0.0154	0.00669	0.2016	0.00007	0.00633	108.3
77	0.14049	0.00247	0.0162	0.00071	0.01017	63.7

[a] Calculated by 0.693/(k_{43} + k_{53}). The value for CB 18 also included a "loss constant" presumed to be volatilization loss.

As indicated in **Chapter 6**, the CBs are differentially metabolized to more polar products. CB 18 yields at least 13 distinct, but unidentified, metabolites retained in the fly and 10 metabolites are resolvable in the excrement (Saghir et al., 1993; Saghir and Hansen, 1992). CB 77 is metabolized more slowly to 7 distinct metabolites in the fly and 8 in the excrement, while the persistent CB 47 yields even smaller amounts of fewer metabolites (Saghir et al., 1994).

Compartment 4 includes unextractable CBs and/or metabolites. A portion of the radioactivity is not extractable from the fly or from the filter paper used to collect the excrement; these fractions are considered to be covalently bound to macromolecules (e.g., Morales and Matthews, 1979; Shimada and Sato, 1978; Shimada and Sawabe, 1984; Saghir et al., 1999). Rate constant, k_{41}, describes this saturable process in the cuticle while k_{43} describes the process in the internal milieu of the fly. The unextractable pools within the fly and the integument were not separated, but a greater proportion of binding occurs via the integument pathway. As with rats (Shimada and Sawabe, 1984; Saghir et al., 1999), more CB 77 is converted to unextractable products than is CB 47 (Saghir et al., 1994).

7.6. Toxicokinetics and PCB Residue Profiles

The experimentally determined toxicokinetic parameters are useful for predicting species- and congener-specific bioaccumulation and retention of PCBs. For the few congeners studied, it is apparent that the more readily metabolized CBs are the least persistent. Other factors such as tissue perfusion rates, absorption routes, lipophilicity of the CB and lipid composition of the toxicokinetic compartment also influence relative congener disposition. Still, species differences in metabolism and factors which influence PCB metabolism such as induction and inhibition of CYPs and phase 2 enzymes (**Section 5.3; Chapter 6**) appear to have the greatest selective effect on individual congeners and, thus, on net PCB accumulation.

The final **Chapter 8** considers net total PCB accumulations and profile changes in food chains. The distinct occurrence, movement, degradation and toxicokinetics of individual PCB congeners as described in the preceding chapters is consistent with the residue profiles observed at different trophic levels. Although precise predictions have not yet been developed, the observations are sufficient to project exposure levels and profiles for more accurate risk assessment.

8 FOOD CHAIN PROFILES

Several examples of food chain changes in PCB residue composition have been introduced in **Chapter 5**. Factors favoring accumulation of specific congeners have been interspersed throughout the text. These factors include air transport and herbage uptake (**Chapter 4**) and air ↔ water ↔ sediment exchanges as they influence the accumulation by aquatic and marine biota (**Chapter 5**). Individual organism metabolism and toxicokinetics were discussed in **Chapters 5, 6** and **7**. Aquatic and marine biota bioconcentrate PCBs and other POPs through contact with sediments and waters as well as through food chains. Bioaccumulation and biomagnification through dietary sources, however, are more important for terrestrial organisms. Each dietary source has a unique profile of specific PCB congeners. For example, the CBs and other POPs present in fish influence the profiles accumulated in fish-eating birds. Birds feeding mainly on invertebrates retain a somewhat different mixture, including a CB 101/118 ratio higher than most other homeotherms (**Section 7.5.2**). PCB-exposed insects, at least, also transfer complex mixtures of retained metabolites (**Chapter 6; Section 7.5.3**). Perhaps of greater importance, certain observed responses in bird populations have been correlated with diet and the CAs dominating the burdens of the different food types (Arenal, 1997; Fox et al., 1997; Fry, 1995; Kubiak et al., 1989). Omnivores accumulate a more balanced mixture of toxicants and may be only briefly and sporadically challenged. Fish have long been considered to provide the greatest food-borne toxicant challenge to omnivores.

8.1. PCB profiles in fish

It is well-recognized that fish and seafood are major routes for entry of environmental POPs into food chains. PCB residue profiles in fish are also recognized as unique from those in birds and mammals. Some profiles have been compared in detail (**Table 5.9; Appendix Table 10; Figures 5.1** and **5.2**). The sample residues from Great Lakes fish in **Appendix Table 10** have been supplemented with four additional relatively complete residue reports to estimate an "average fish" profile in **Appendix Table 11**. Note that there are unequal (1-9) entries for each CB and the "percent" in each profile is heavily influenced by the numbers of analytes; thus, the "average" profile is probably not at all characteristic of any one fish or species. The two tables do, however, highlight the unique nature of fish PCB profiles.

Marine mammals also have somewhat unique PCB profiles, due partly to diets related to fish or fish-feeders. Humans and other omnivores reach steady state POP body burdens through a varied diet. Brief deviations from steady state profiles are likely to occur following certain components of this diet, such as fish. The characteristic PCB compositions in mammals which are more consistent fish consumers may help to predict the influences of fish meals on human profiles.

8.2. PCB profiles in marine mammals

Seals provide a suitable model representing a more consistent pattern of fish consumption than do omnivorous mammals. There are species-specific variations in PCB biodegradation capacities, but marine mammals generally degrade PCBs according to SARs similar to those for other mammals (e.g., Boon et al., 1994; 1997; Bruhn et al., 1995; Letcher et al., 1997; Muir et al., 1988; Norstrom et al., 1988; 1992). Reported profiles of PCB congeners in the blubber of seals reflect a fish-based diet (**Table 8.1**), but the patterns are not uniform. The variable methods and selections of analytes permit a broader perspective for CBs detected, but the "percents" are dependent on the numbers of analytes reported.

Table 8.1. Estimated proportions (% reported congeners) of PCB congeners in the blubber of seals from various locations.

B & Z No.	Ringed n = 33[a]	Harbor n = 7[b]	Harbor n = 9[c]	Harbor n = 7[d]	Caspian n = 1[e]
8			0.01		
17 + 18	<0.2				
18			0.03		
28 + 31	1.0	0.10	0.23	0.3	0.03
37		0.0001			
40	<0.2				
44	<0.2		0.25	0.7	0
47 (+ 48)	1.0				0
49	1.0				0.15
52	4.5	1.44	3.20	6.2	0.34
56 (+ 60)	1.6				0.02
60					0.01
66 (+ 95)	1.2		0.44		0.04
70 (+ 76)	0.6	0.07			0.09

B & Z No.	Ringed n = 33[a]	Harbor n = 7[b]	Harbor n = 9[c]	Harbor n = 7[d]	Caspian n = 1[e]
74	1.9				0.13
77		0.0002	0.001	0.002	0.005
81		0.0001			
84	0.8				0
85	0.8				1.46
87	0.8				0.17
92					0.32
95 (w 66)		0.89			0.39
97	<0.2				0
99	7.2				7.02
101 + 90	9.7	3.98	7.51	2.0	1.42
105 (+ 132)	2.1	0.64	1.23	1.7	0.21
107					0.04
110 (+77)	1.4	0.41		5.5	0.25
114	0.4				
118 (+ 149)	6.2	1.70	4.19	1.9	0.70
126		0.004	0.008	0.016	0.014
128 + 167	1.4	3.54	3.66	2.6	2.83
132 (w 105)				3.9	0.51
135					0.23
137	0.8	0.46			
138 (+ 163)	16.0	28.47	24.77	17.8	19.89
141	<0.2				0.24
146	3.1				3.21
149 (w 118)	2.7	9.88		2.8	2.95
151	0.4				0.18
153	19.5	35.44	28.22	22.2	29.00
156 (+ 171)	0.8	0.66		1.6	0.46
157	0.2	0.22			0.15
158		1.10			2.00
163 (w 138)				2.1	6.26
167 (w 128)		0.16			0.015

B & Z No.	Ringed n = 33[a]	Harbor n = 7[b]	Harbor n = 9[c]	Harbor n = 7[d]	Caspian n = 1[e]
169		0.0003	0.0002	0.0002	0.002
170 + 190	1.4	2.98	3.99	3.8	2.45
171					0.36
172 (+ 192)	0.2				
174	0.4				0.23
177	0.2				1.27
178	1.0				1.06
180	4.5	7.32	10.70	7.1	7.08
183	1.6			2.3	1.95
185					0
187	2.9		9.02	8.2	3.31
189		0.07			0.10
190 (w 170)				1.2	0.49
194	0.2	0.56		1.2	0.52
195			0.84		
199	0.6				0.46
201	<0.2				
203 + 196	0.2				
206	<0.2		1.18		
209	<0.2		0.51		

[a] Arctic ringed seals (*Phoca hispida*), Bering Strait, April, 1984 (Muir et al., 1988).
[b] Danish harbour seals (*Phoca vitulina*), Kattegat Sea, 1988 (Storr-Hansen, 1992; Storr-Hansen and Spliid, 1993).
[c] Harbor seals, Northeast coast of United States, 1990-1992 (Lake et al., 1995).
[d] Harbour seal pups, dead or moribund, sampled within 12 hours of death, Norfolk Wash, UK, 1988 (Troisi and Mason, 1997).
[e] Single male Caspian seal (*Phoca caspica*) found dead near Iran, December, 1993 (Vetter et al., 1995).

CBs 138 and 153 dominate the PCB congeners in the seals as with most other biota. In two cases, CB 163 was resolved from the CB 138 peak and found to constitute 10-24% of the peak. CB 132 accounted for 70% of the CB 105/132 peak in both cases, resulting in a relatively low level of mono-*ortho* CB 105. Note that in many other cases, the contribution of CB 105 may have been greatly overestimated. It accounted for only 20, 38 and 16% of the total CB 105 and CB 132 in flounder, herring and porpoise residues, respectively (Kannan et al., 1995; **Table 5.9**). In some cases, mono-*ortho* CB 118 also made a smaller contribution than in other mammals, but TCDD-EQs were more than compensated by the consistently high levels of the highly potent CB 126. The labile CBs 52, 101 and 149 averaged greater relative proportions than in other mammalian residues, probably reflecting the profiles in fish. CB 110 occurred at lower average proportions than in fish. In seals from the Wadden Sea, only 1/7 had detectable residues of CB 110 (Storr-Hansen, 1992). This is consistent with the rapid metabolism of this CB by other mammals (**Chapters 3 and 7**).

Detailed examination of profiles can permit identification of seals from the same colonies and distinguish seals of the same species from other locations (Storr-Hansen and Spliid, 1993). The congener profiles also differ in Arctic locations as well as with age, which may be a reflection of diet (Muir et al., 1988). Walrus which feed primarily at lower trophic levels (e.g. mussels) accumulate a broader profile of CBs, including lower chlorinated homologs, than those feeding at higher trophic levels, presumed to include seals (Muir et al., 1995). Likewise, polar bears accumulate a PCB profile more dominated by the major persistent congeners (CBs 99, 138, 153, 180) (Muir et al., 1988; Norstrom et al., 1988).

These variations in residue profiles among consistently marine-oriented carnivores suggest that omnivorous mammals will demonstrate even greater variations in PCB profiles. The quantitative differences may, however, be smaller because of the averaging effect of several sources. The changes may also be quite sporadic. Additional sources of PCBs include airborne exposures, both directly and indirectly through accumulation in herbage. These routes are generally not important in seals, but may predict more complex alterations in humans.

8.3. Other profile-altering PCB exposures

The characteristic PCB compositions of exposure pathways other than fish consumption have also been estimated from various reports as a preliminary step in developing the ability to predict pulsatile deviations in PCB profiles for omnivorous mammals. Sample air, herbage and total human diet PCB compositions were developed in a manner similar to the development of "average" fish PCB compositions in **Appendix Tables 10** and **11**. Again, only selected sources reporting broader selections of analytes were consulted. Average relative production estimates (**Table 2.6; Appendix Table 4**) are compared to preliminary average percent compositions of PCB residues in air, fish, produce and a total diet in **Table 8.2**.

Table 8.2. Proportions of chlorobiphenyls in various exposure media compared to estimated production figures.

B&Z No	Structure of first-listed	Relative Production[a]	Congener Proportions[b]			
			Air	Fish	Plant[c]	Diet[d]
1	2	0.5	0.6			
3	4	0.2	0.6			
4	2-2	1.3	0.9	+		
6	2-3	0.6	1.0		0.7	
8	2-4	3.0	0.3	0.5	2.9	
15	4-4	0.9	2.2	+		
16,32	23-2	1.4	6.6	0.4	1.2	
17	24-2	1.4	2.9	0.2		
18	25-2	4.0	5.9	0.4	2.9	6.0
19	26-2	0.4	1.5			
22	23-4	1.4	4.7	+	1.6	
25	24-3	0.2	0.3	0.3		
26	25-3	0.6	1.8			
28	24-4	3.5	16.7	1.0	5.0	5.0
31	25-4	3.8	2.8	0.7	3.7	4.8
32	26-4	0.9	?	?	?	
33	34-2	2.4	3.2	0.5	1.8	
37(42)	34-4	1.0	2.8	?	1.8	
40	23-23	0.5	1.3	0.5		+
42(37)	23-24	0.8	2.9	1.6		
44	23-25	2.7	4.6	1.5	2.0	2.4
45	236-2	0.5	2.1	+		
46	23-26	0.2	0.5	+		
47,48	24-24	0.7	2.8	2.8	+	1.6
48	245-2	0.7	?	?	?	?
49	24-25	1.8	3.2	2.5	2.4	1.7
51	24-26	0.1	0.4	+		
52	25-25	3.5	6.1	1.9	4.6	5.0

B&Z No	Structure of first-listed	Relative Production[a]	Congener Proportions[b]			
			Air	Fish	Plant[c]	Diet[d]
56,60	23-34	1.3	2.2	1.8		1.0
60	234-4	1.4	?	?		?
64	236-4	1.3	1.8	1.6		
66	24-34	2.5	4.5	3.2	1.8	2.5
70	25-34	3.3	3.5	1.8	2.6	3.6
71	26-34	0.7				
74	245-4	1.5	1.5	1.1	0.9	2.6
77,110	34-34	0.2	0.2	0.2	+	0.4
81	345-4	0.01	0.01	(+)		
82	234-23	0.4	0.4	+		
84	236-23	0.8	0.8	3.8	1.8	
85	234-24	0.4	0.4	1.8		
87	234-25	1.3	1.3	3.1	1.7	1.8
91	236-24	0.4	0.5	1.5		
92	235-25	0.4	0.5			
95	236-25	2.3	1.8	2.4	5.2	4.4
97	245-23	0.8	0.7	1.5	2.1	1.2
99	245-24	1.1	0.4	3.6	2.2	2.2
101	245-25	2.9	2.5	4.3	4.1	5.6
105,132	234-34	1.1	0.2	2.9?[e]	0.9	1.6
110,77	236-34	3.0	2.0	5.0	9.0	4.3
114,134	2345-4	0.07	0.01			0.03
118	245-34	2.2	0.06	5.4	3.2	4.7
119	246-34	0.02	0.05			
123	345-24	0.05				0.04
126	345-34	0.001	0.002	+		0.04
128	234-234	0.4	0.1	2.0	1.1	?
132	234-236	1.2	0.1	2.0?[e]	?	?
135	235-236	0.4	0.4		1.5	
136	236-236	0.5	0.4	0.9	1.1	1.7

B&Z No	Structure of first-listed	Relative Production[a]	Congener Proportions[b]			
			Air	Fish	Plant[c]	Diet[d]
137	2345-24	0.1		0.3		1.3
138,163	234-245	2.8	0.8	7.4	3.7	12.1
141	2345-25	0.8	0.2	1.3	0.8	1.6
146	235-245	0.4	0.2	1.6		
149	236-245	2.8	0.7	3.6	3.9	5.0
151	2356-25	0.8	0.3	1.3	2.2	2.1
153	245-245	3.0	1.4	8.6	6.2	13.0
156	2345-34	0.3	0.15	1.3	0.3	1.0
157	234-345	0.04	0.03		0.04	0.1
158	2346-34	0.3	0.05	+		
163	2356-34	0.8	?	?	?	?
167	245-345	0.1		+	0.3	0.9
169	345-345	<0.001	0.001			
170,190	2345-234	1.0	0.3	2.2	1.1	2.3
171	2346-234	0.2		0.6	0.5	0.6
174	2345-236	1.2	0.3	1.6	0.8	1.7
177	2356-234	0.6	0.1	1.0		0.5
179	2356-236	0.5		0.6		
180	2345-245	2.8	0.4	3.8	0.8	4.4
182	2345-246	0				
183	2346-245	0.6	0.3	1.4	0.3	1.4
187	2356-245	1.4	0.6	2.6	0.5	2.4
189	2345-345	0.02	T		0.06	0.1
190,170	23456-34	0.2	?	?		
194	2345-2345	0.5	0.05	0.4	+	
195	23456-234	0.2	0.3		0.5	
196,203	2345-2346	0.3	0.4	0.9	0.2	
199	2345-2356	0.4	0.3	0.9	0.1	
203,196	23456-245	0.4	?	?	?	
206	23456-2345	0.9		0.1	0.4	

B&Z No	Structure of first-listed	Relative Production[a]	Congener Proportions[b]			
			Air	Fish	Plant[c]	Diet[d]
209	23456-23456			0.05		

[a] Estimated relative amount produced from **Appendix Table 4**.
[b] Approximate averages of several reports which included a broad selection of analytes. Fish from **Appendix Tables 10** and **11**; Air from **Appendix Table 6**.
[c] "Plants" refer to edible produce (tomatoes from **Appendix Table 7** supplemented with additional data from Cullen et al., 1996). Data were averaged from 23 tomato, 17 potato, 11 carrot and 9 lettuce samples from different locations (see text).
[d] The "Diets" are averaged from the various regions of Italy; these represent the most complete congener profiles located (Turrio-Baldassarri et al., 1997; 1998; Zuccato et al., 1999).
[e] When not resolved, it is likely that CB 105 constitutes only 15-40% of the peak in fish (**Tables 5.9; 8.1**).

The produce samples were variously obtained from the New Bedford, MA area. Although from a single study, the plant residues represent a variety of conditions and are enriched in steady state congeners due to proximity to a Superfund site. Samples from near the contaminated site contained greater proportions of more heavily chlorinated homologs than did samples distant from the site (non-local). Tomato samples taken downwind during dredging had higher proportions of mid-eluting congeners (Cullen et al., 1996). The shift in profile close to contamination was similar to that seen in vegetation downwind from a transformer furnace leak (**Section 4.5, 4.6**; Froese et al., 1998). Thus, produce grown in ambient PCB conditions would deviate even more markedly in the proportions of "major" CBs.

It is known that there are subtle differences in the residue composition of poultry and eggs, beef, lamb and pork, and even milk and cheese (Hansen, 1979; 1987; Turrio-Baldassarri et al., 1995; 1996; 1997; 1998). For simplicity, however, humans will be considered as representative homeotherms and the "average" PCB residue compositions in meat, milk, eggs and poultry will be considered to be similar to those in humans. The effect of exposure via these similar-profile-to-human food sources, in cases where residues are higher than ambient, would be to elevate the levels of the steady state profiles in humans. This would further mask sporadic exposures to more human-labile congeners through air, produce and fish by reducing their net contribution. Nevertheless, such transient alterations of steady state PCB profiles do occur.

Pulsatile exposures may have unique toxicological implications. The understandings of the biological actions of PCBs which act through the Ah receptor are well-characterized. The understanding of the biological actions of pulsatile congeners and/or their metabolites is still not adequate to assess potential risks. **Table 8.2** and a more flexible model of exposure presented later may help to prioritize congeners for further study. Should a biologically plausible risk be suggested, this framework may be further useful for predicting and modifying exposures, especially for more susceptible developing animals. These data have been compiled for initial guidance and cannot be taken literally until more complete comparisons are possible.

Data generation requires little additional effort by the analyst (Hess et al., 1995). Data analysis requires considerable additional effort (e.g., Devoto et al., 1997) and an increased level of complexity inserted into risk analysis, so it is necessary to provide some rationale for the need.

8.3.1. Public Health Implications

If adverse medical outcomes from PCB exposures occur, they are most likely associated with pulses of exposure in reproducing and developing organisms (Hansen, 1998). Adult human PCB exposures and steady state body burdens have generally failed to be consistently associated with serious adverse medical outcomes, even in former occupationally exposed groups. The ambiguity of results may be partly due to failure to consider the complete profile of PCB congeners and other POPs present. Even with a comprehensive list of analytes, failure to detect (or notice) unique profiles due to pulses of exposure may also lead to ambiguous associations. Because of the ubiquity of PCBs, these ambiguities are likely to suggest false positive associations with health effects as well as false negatives.

Many recent studies have focused on the dioxin equivalents in human milk and/or correlations with developmental outcomes (e.g., Dewailly et al., 1991; Koopman-Esseboom et al., 1994; 1996). Generally, coplanar and mono-*ortho* PCBs contribute the greatest TCDD EQs in human milk and correlations with developmental outcomes are variable. The contributions of PCBs to TEQs are important considerations because CBs 77 and 126 and, especially, the mono-*ortho* coplanar congeners have additional biological actions independent of the Ah receptor and TCDD EQs (Hansen, 1998; **Table 2.7**).

This raises the question: If the correlations of adverse developmental outcomes with TCDD EQs are weak, could they be strengthened by considering correlations with, e.g., phenobarbital equivalents since the mono-*ortho* congeners contribute strongly to both types of actions? That is, the overwhelming contribution of PCBs to TCDD EQs may cause a coincidental weak correlation because of Ah Receptor-independent properties of the same PCBs; therefore, the correlation might be stronger if some exposure parameter other than TCDD EQ or total PCB were used.

This question cannot be answered at present. It is hoped that the thorough characterization of exposure profiles will permit selection of significant congeners and/or metabolites for the development of additional biological action profiles or equivalencies similar to those for TCDD EQs (Wolff et al., 1997; Hansen, 1998). Phenobarbital EQs may be used to predict vitamin K deficiency in the neonate (Bouwman et al., 1992; Canfield et al., 1995) while high TCDD EQs may suggest vitamin A supplementation. An emerging equivalency of possible endocrine, neurochemical and immunological significance is associated with the ryanodine-sensitive calcium channels via ryanodine receptors (RyR_1, RyR_2, etc.). Tri-*ortho* and di-*ortho* CBs are the most active while non-*ortho* and mono-*ortho* CBs are essentially inactive (Wong and Pessah, 1996; 1997; Schantz et al., 1997). The preliminary SARs were comparable to those for other AhR-independent neurochemical effects (see Hansen, 1998), but more complete SARs (Pessah et al., 1999) and descriptions of mechanisms are emerging.

It has already been demonstrated that pulses of exposure (2 days) to environmental PCBs and even labile congeners can have significant enzyme and endocrine effects in immature rats (e.g., Hansen, 1998; Li and Hansen, 1996; Li et al. 1994; 1998). Weanling swine receiving 15 mg/kg total Aroclor 1254 over 1 week (about 300 mg) suffered reduced weight gains over the entire 4-5 month observation period (Hansen, 1979; Hansen and Welborn, 1977). Weanling swine receiving 20 ppm Aroclors 1242 and 1254 in the diet for 3 months (5000 mg) suffered similar reduced gains (Hansen et al., 1977). Relative liver weights were not changed in the 3-month dietary exposure or 1 week of CB 153, but were significantly increased by 1 week of CB 52, CB 118 or Aroclor 1254. Relative thyroid weights were slightly lower in the acutely dosed pigs than in the diet-exposed pigs. The relative adrenal weights were significantly increased at 3 months in pigs having received 5000 mg Aroclors 1242 or 1254 in the diet; however, after 5 months, the relative adrenal weights in pigs having received 300 mg CB 118 during the first week after weaning were increased to a greater extent.

The events initiating actions leading to eventual adverse medical outcomes may occur within a brief window of opportunity. This would be especially true in developing humans and other animals because there are more windows of opportunity, the susceptible processes are generally more sensitive to insult, and the lag time between event and manifestation may be longer. By the time the adverse outcome is apparent, the PCB (and other toxicant) profiles will no longer resemble the profile present when the events occurred (Hansen, 1998; vom Saal et al., 1998).

It may be of note that in the National Human Adipose Tissue Survey (NHATS), infants and children (ages 0-14) had detectable levels of triCBs but no nonaCBs (Phillips and Birchard, 1991). TriCBs were nondetectable in age 15-44, but nonaCBs were detected. Even if labile CBs are below detection limits, higher residues of persistent congeners often implies the presence of labile congener metabolites. Hydroxy CBs are often more potent estrogens and thyroid hormone depleters (Korach et al., 1989; Lans et al., 1993; 1994). Catechol and quinone products are also likely, especially for the less chlorinated homologs (**Chapter 6**) and may be even more biologically active (e.g., McLean et al., 1996a,b). Persistent hydroxy metabolites are cytotoxic and/or antiestrogens; however, they persist bound to plasma proteins and may exert additional imbalances in this way or, perhaps, just be preserved as inertly bound species (Bergman et al., 1994; Kramer et al., 1997; Hansen, 1998).

The 236 and 25 substitutions which occur in many labile congeners favor production of methylsulfonyl metabolites (**Chapter 6**). Labile congeners such as CBs 52, 101, 132 and 149 may not be found at high levels in average humans, but MeSO metabolites are significant components of the total burden. Some 3-methylsulfonyl PCBs are potent phenobarbital type inducers (Kato et al., 1995; 1997), while some 3- and especially 4-methylsulfonyl PCBs interfere with glucocorticoid receptors (Johansson et al., 1998).

Blood PCB levels were highly correlated with hormone and immune abnormalities in women with repeated miscarriages in Germany (Gerhard et al., 1998). Since total PCB was estimated from the sum of CBs 101 + 138 + 153 + 180, it is not possible to determine if associations were with persistent parent CBs or coincident metabolites of CB 101 and other labile congeners.

Weak correlations between other adverse medical outcomes and average exposure categories may well be due to misclassification of exposure following data reduction necessitated by the large numbers of analytes. The default categorization of subjects is based on recognized patterns and/or abridged selection of analytes and/or the availability of convenient additive categories such as TCDD EQs. "Average" exposures and body burdens probably represent very few, if any, actual exposures and body burdens. The landmark studies of PCB-associated neurotoxicity (e.g., Jacobson and Jacobson, 1996; Koopman-Esseboom et al., 1996; Gladen and Rogan, 1991) used broad exposure categories or limited residue profiles for subject classifications. The controversial breast cancer studies (Wolff et al., 1993; Krieger et al., 1994; Hunter et al., 1997) used aged steady state total PCB (or higher homolog only) values (Wolff et al., 1997).

Consideration of the emerging reports of the biological actions of metabolites and specific CBs, other than AhR agonists, suggests some correlations may be demonstrable. Examples of more detailed consideration of individual analytes illustrate the deviations of congener occurrence within human populations and the enhanced resolution of more specific biologically plausible relationships. With small numbers of subjects, Dewailley et al. (1994) were able to show a positive correlation between the weakly estrogenic PCB 99 and DDE and estrogen receptor positive breast cancer. The putative antiestrogenic PCB 118 was significantly lower in patients with estrogen receptor negative breast cancers.

Hardell et al. (1996) found no correlation between DDE levels and risk for non-Hodgkins lymphoma (NHL). There were, however, strong NHL risk correlations for several more heavily chlorinated *ortho*-rich CBs and mono-*ortho* CBs (156, 157). There were no correlations with coplanar congeners, PCDDs, PCDFs or TCDD-EQs (Hardell et al., 1996). Using nested case-controls and measurement of 28 CBs as well as several pesticides, another high correlation between serum PCB and risk of NHL was demonstrated (Rothman et al., 1997). The mean time to diagnosis was 12.1 $\pm$ 5.2 years after the serum sample, so the residues were more likely to represent the profile present during the insults. Again, there were no correlations with NHL for any of several pesticides.

In order to accurately determine whether the modest PCB and TCDD exposures experienced with declining global loads involve real health risks, it is necessary to examine the exposures more accurately and in more detail. Data reduction and analytical as well as statistical considerations may introduce a false "tendency for homogeneity" (DeVoto et al., 1997) or consistency (Duarte-Davidson et al., 1994) in PCB profiles. Until some readily applicable statistical and population-based methods are developed to correctly categorize subjects, there is little hope for the accurate risk assessment of complex mixtures of POPs. Accurate risk assessment may also exonerate PCBs of some suspected toxicities so the search for the real cause(s) can focus on other agents.

8.4. Reporting of Human PCB Residues

As with other biotic systems, the primary determinant of PCB and other POP body burdens is exposure. Steady state residues reflect multiple pathway exposures and net elimination. PCB profiles in human serum immediately following exposures reflect the profiles in the exposure sources (**Chapter 7**); however, selective metabolism and excretion begin altering the profile within 4 - 24 hours. Pulses of other-than-ambient exposures are generally not reflected in surveys of populations because of infrequent occurrence. If abnormal profiles do occur in a small proportion of a surveyed population, they are lost in the data reduction necessary to comprehend the general trends. The regular occurrence of "abnormal" PCB profiles will be demonstrated by reversing this data reduction to arrive at actual levels of low-frequency congeners in certain individuals.

It was demonstrated in **Chapter 3** that expanded specific congener analysis of human residues results in a very unmanageable data set (**Table 3.3; Appendix Table 5**). Efforts to reduce the data to readily intelligible trends frequently discard information that may be trivial to a population but critical to the individual. Surveys tend to mask low-frequency PCB profiles by averaging all subjects and/or pooling samples. Closer examination, however, can still detect unique exposures as well as composition changes. There is an encouraging trend to examine outliers more carefully and even consider them for medical surveillance or intervention (e.g. Korrick and Altshul, 1998; Humphrey et al., 1999).

Total dietary studies from Italy have recently provided some of the broader profiles of PCB exposures (**Table 8.2**). The often surprising high levels of congeners not usually associated with steady state human residues confirms probable exposures to unique PCB profiles. The human residues, however, do not reflect these exposures when pooled samples are analyzed (Turrio-Baldassarri et al., 1996). Likewise, data from analyte-rich surveys (e.g. Duarte-Davidson et al., 1994; Mes et al., 1993; Newsome et al., 1995; Vartiainen et al., 1997; Anderson et al., 1998) must be reduced to a manageable set for comprehension. Statistical comparisons reveal, e.g., regional, age, and parity differences in average POP contents of human serum and breast milk. If the individual data are examined, however, additional sometimes profound trends can be observed (Whittem et al., 1998). Nevertheless, when dealing with more complete lists of analytes even very large data bases do not provide the statistical power to resolve potentially medically important regional and subject differences in PCB profiles.

For example, the survey of human breast milk in Canada (Newsome et al., 1995) reported mean levels for 40 PCB congeners, 3 DDT derivatives, 6 chlorobenzenes, and a dozen other POPs. A total of 497 breast milk samples were analyzed from throughout Canada and a trend in declining POP burdens from an earlier survey (Mes et al., 1993) was detected. Mean levels of individual analytes could be compared, but ranges were broad and some analytes (e.g., α and γ HCH, 1234 and 1235 tetrachlorobenzenes, PCBs 33, 37, 52 and 110) were infrequently detected. In spite of author explanations, the effect of the necessary data reduction was to *permit* misinterpretation as rather uniformly distributed burdens of these analytes at mean and/or median levels well-below their actual occurrence in several individuals.

Figure 8.1 presents modified mean values (average of positive samples only) for several PCB congeners as well as *o,p*-DDE and *p,p*-DDT. Even then the relative SDs for universally present CBs 118, 138, 153 and 180 are in excess of 50%. The frequently detected PCB 66 was 3 X the level of CB 52 in overall mean milk content (Newsome et al., 1995); however, when only positive samples are averaged, contents are nearly equal (**Figure 8.1**). Three individuals were selected by id number without regard to demographics or analytes detected. The numbers selected were 111, 222, and 333; 222 was a nonresponder so a default number, 22, was substituted. All were in their 20s, in mid-lactation, and nursing their first or second child. None of these individual milk samples had detectable levels of the highly variable CB 44 and one had an extraordinarily high level of CB 37 (+ CB 42?) (**Figure 8.1**). In addition, 2 had levels of CB 52 consistent with the mean of positive samples but well in excess of the mean of all samples. Only 1 had a detectable level of the endocrine active (Li et al., 1998) CB 110 which was double the mean of positive samples and 10 X higher than the mean of all samples. When

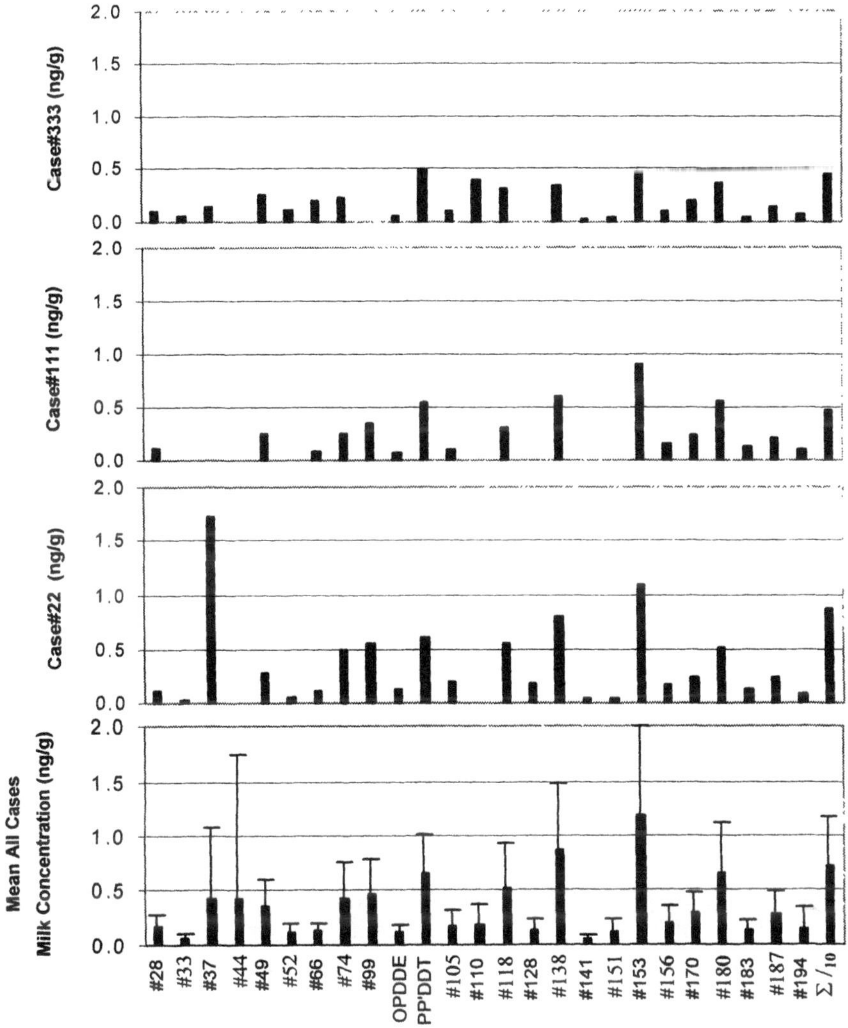

Figure 8.1. Mean ± SD concentrations of selected CBs in human breast milk samples from Canada (bottom). Means derived from positive samples only in data provided by W.H. Newsome. Top 3 profiles are from individual samples selected by number without regard to demographics. Total PCB (far right) is divided by 10 to maintain scale.

all POPs analyzed are factored into these comparisons, it is **inconceivable** that the biological challenges to the nursing infants of these three mothers would be similar. Nevertheless, they had similar total PCBs and total TEQs so would be placed in similar categories if they were subjects of epidemiological studies searching for adverse medical outcomes.

8.4.1. Frequencies of detection and impact on conceptualization

Those CBs which occur in fewer individual samples are easily discounted by averaging the entire data set (including non-detect individuals) or deleting analytes which occur at less than a reasonable minimum frequency. These data reduction techniques are acceptable, and perhaps necessary, for discerning broad temporal and population trends. On the other hand, they are not acceptable when applied to epidemiological studies or specific risk analyses because they may obscure real relationships. In addition, if only the "major" CBs (e.g., 99, 118, 138, 153, 180) are reported, they are often present in similar proportions and would not reveal deviations from steady state.

A simple exercise utilizing a (rare) complete published data set can emphasize the effect of data reduction. In a study comparing fish and lobster PCBs with human serum PCBs, the analytical results for 23 individuals were reported (Burse et al., 1994; **Table 3.2**). Two had very unusual (occupational?) exposures and should be examined separately, but the remaining 21 subjects could be relatively easily transferred to a spread sheet. Except for CBs 52, 101 and 177, the modified data reduction (**Table 8.3**) still results in mean values within a factor of 2 whether means are derived by averaging zero values or only positive samples. Yet, even in this small data set with extreme outliers removed, there are unique profiles.

The dramatic increase in actual vs average CB 52 concentrations in serum from a small proportion of subjects was also reflected in the Canadian breast milk data which had a much higher *n* (**Figure 8.1**). CB 101 was similarly detected in only about half of the Canadian samples. Breast milk appears to be a less selective reservoir for PCBs, retaining higher proportions of more labile congeners, than are serum or adipose tissue (Bush et al., 1985; Duarte-Davidson et al., 1994; Hansen, 1998). Occupationally exposed humans, however, also retain some labile congeners due to more recent exposure and continued exposure to work-related reservoirs (**Appendix Table 5; Tables 7.6** and **7.7**). The rather uniform decline in PCB residues in children (**Table 7.7**) and the unusual contribution of triCBs to age 0-14 NHATS samples (Phillips and Birchard, 1991) suggests that children dilute PCBs by growth rather than eliminate them by metabolism.

A summary table using multiple analyte-rich reports confirms that human breast milk can contain significant proportions of rather labile congeners in addition to those most often reported (**Appendix Table 12**). Pattern recognition techniques are useful for determining sources of contamination (e.g., Anderson et al., 1998; DeVoto et al., 1997; Luotamo, 1991; MacDonald et al., 1992; Needham et al., 1996). At the same time, using such techniques discounts sporadic exposures which may be of public health significance. Many "pulsatile" congeners are generally not considered in patterns due to low frequencies and/or omission from the target analytes. Some of the congeners which may escape notice in specific surveys are suggested in **Appendix Table 13** which presents frequencies above detection limits.

Detection limits play a critical role in the numerous reports which summarize the major congeners found to be representative of the PCB residues examined. Reasonable quantitative estimates can be derived from a few analytes and correlations can be quite high. Anderson et al. (1998) confirmed anglers as reasonable sentinels of sports fish contaminants and suggestive of contaminants to monitor. They measured 32 CBs + the 4 coplanar CBs and found 19 to be present in all samples from Lake Michigan, Lake Huron and Lake Erie fishermen. Maack and Sonzogni (1988) and Niimi and Oliver (1989) found the more labile congeners (CBs 66/95, 101, 110) to be most consistently present in fish from the Great Lakes Basin, but these were not common in the serum of Wisconsin sports fish consumers (Sonzogni et al., 1991). The latter study reported a detection limit of 0.6 ng/g and the frequency of CBs 44 and 52 were "0"; however, the frequencies of detection for CBs 138, 153 and 180 were also below 80% in this study. DeVoto et al. (1997) acknowledged unusual PCB exposures may result in atypical profiles, but chose to report only congeners which occurred in more than 50% of the samples and limit correlation analysis to those quantifiable in 90% or more of the samples. Using these restrictions, it was determined that quantitation of a few congeners could "yield essentially the same information as measurements of a large panel, and be more cost efficient."

Again, these techniques are appropriate for discerning trends, but are inappropriate for epidemiological classifications and/or medical surveillance. In spite of the high detection limits, Sonzogni et al. (1991) found 15/173 subjects with detectable CBs 74, 99, 146, 156, 183, 194, 201 and 203. The 13 congeners suggested by Sonzogni et al. (1991) accounted for only 64% of the total serum PCB in 101 fisheaters when 22 congeners were used (Humphrey et al. 1999). These 22 congeners accounted for greater than 90% of total PCB in the serum of most of the 101 fisheaters and 78 non-fisheaters (Humphrey et al., 1999). Many of the unique subjects had CB 138/163 levels greater than those of CB 153; the commonly encountered sporadic peaks included CBs 66/95, 101, and 110, with CBs 56, 87, 97, 151, 171 and 183 being less common (Humphrey et al., personal communication). Infrequent CBs

Table 8.3. Frequency and profile of CBs in human serum from fish consumers near the New Bedford, MA area (Adapted from Burse et al., 1994).

B & Z No	Positive Samples	Serum (ng/g)		
		SUM of congener	Mean of Positive	Overall Mean
28	18	229.96	12.78	10.95
52	3	2.87	0.96	0.14
56 + 60	15	30.71	2.05	1.46
66 + 95	17	25.36	1.49	1.21
74	21	189.73		9.03
99	21	99.84		4.75
101	10	5.46	0.55	0.26
105	20	56.52	2.83	2.69
118	21	98.06		4.67
130[a]	20	24.41	1.22	1.16
131[a]	16	18.52	1.16	0.88
137	21	33.99		1.62
138 + 163	21	289.30		13.78
146	21	168.36		8.02
153 + 132	21	322.98		15.38
156	21	98.36		4.68
167	21	27.81		1.32
170	21	56.00		2.67
177	8	4.53	0.57	0.22
180	21	154.52		7.36
183	17	19.62	1.15	0.93
187 + 182	21	53.65		2.55
194	19	16.98	0.89	0.81
200 + 157	19	28.81	1.52	1.37
201	20	22.06	1.10	1.05
203 + 196	14	13.19	0.94	0.63
206	9	5.38	0.60	0.26
209	18	3.03	0.17	0.14

[a] The high frequencies of CBs 130 and 131 may be unique to New Bedford and similarly dechlorinated residues (**Appendix Table 9**).

were also found in fisheater serum by Burse et al. (1994; **Table 8.3**). Even considering only 11 major congeners, Korrick and Altshul (1998) found unusual profiles in the 4 highest (out of 122) breast milk samples from the same area. All of these observations are consistent with those from the non-uniform congener distributions found in **Figure 8.1** (Newsome et al. 1995). The atypical profiles are typical of sporadic changes in steady state profiles in omnivores. The implications/consequences should be addressed in a more methodical manner.

8.5. Characterizing Human PCB Residues

Also included in **Appendix Table 13** is an additional example of the effect of group averages on interpretation of individual data. For the Welsh breast milk samples (Duarte-Davidson et al., 1994), the high sample is divided by the mean value. The ratio is only 2-4 for the major congeners except that the high CB 99 and CB 118 samples were about 9 times higher than the mean. Individuals had concentrations more than 10 times higher than mean values for CBs 66/95, 77/110, 105/132, 151 and 187 and 28 times higher for CB 149. It is doubtful that the high for each congener occurred in the same sample (e.g., **Figure 8.1**); it is more likely that several individuals experienced sporadic exposures through different routes.

Sporadic accidental exposures to PCBs will continue to occur and these are generally well-monitored when detected. Analytes are selected based on the source. The most constant exposure, however, is to dietary sources.

8.5.1 Dietary Contributions

Certainly the presence of highly potent and additive AhR ligands in the diet is of concern. The more recent studies of human and animal exposures do an admirable job of quantitating the potent trace contaminants contributing to TCDD-like effects. Two decades of detailed studies make the determination of total TCDD-EQs in the diet more accurate than for almost any other group of food-borne toxicants. Unfortunately, the effort required for these analyses and determinations results in neglecting the more prevalent (but less potent and less readily categorized) congeners which cause a totally different spectrum of toxicities. Selected major congeners are also often examined and used as the basis for "total PCB" correlations; in some cases they may be considered individually. Indeed, there was no correlation between TCDD-EQs and risk of NHL, but a high correlation for persistent *ortho*-CBs (**Section 8.3.1**).

Although nearly every current survey of CAs or POPs now includes some measure of TEQs, most of these current studies include very abridged lists of other analytes, especially the complex profiles of PCBs. Fish and seafood have long been known to be important sources of PCBs and TCDD-like compounds and the profiles are known to be unique; thus, published surveys including fish and/or fish consumption are more apt to provide a more complete PCB profile. Many CBs characteristic of seal residues (**Table 8.1**) occur sporadically in humans following fish meals; however, since they are not characteristic of the typical human PCBs, these pulses are seldom apparent in reports (see previous section).

Vegetation, including consumable produce, also accumulates unique profiles of PCBs; the total PCB amounts are much lower than those in fish, but the trend toward fruits and vegetables and away from fish (from the more contaminated areas) as well as fatty meats may increase the relative importance of the major congeners found in produce. For example, human PCB exposure in Viet Nam is higher than most surveyed countries and about 70% of this can occur through herbage-derived foodstuffs because of the high proportions of cereals, fruits and vegetables and the low consumption of fish, meats and oils (Kannan et al., 1992).

In plants, mammalian-labile CBs 8, 18, 101, and 110 are decidedly dominant while CBs 28, 31, 37, 52, 95, 118, 138, 149, and 153 are common (generally 4-8% of reported congeners). Italian regional diets were shown to be dominated by CB 153 followed by CB 138 + 163 (Turrio-Baldissarri et al., 1998). CBs 28, 101, 118, 149 and 180 made about equal contributions to two of the three regions while CBs 18, 52, 141, and 187 were much more dominant in the diet from the central region. CBs 44, 47+48, 49, 66, 74, 99, 110, 151 and 170 made about equal and slightly lower contributions. "Other" probably sporadic congeners were thus mixed rather equivalently with the "Major" congeners other than CBs 153 and 138. These were very large pooled samples, so some components (e.g. produce) would probably have contained unique profiles to account for the large contributions from CBs 18, 52, 101, 110, 141, and 149. Their proportions would have been decidedly higher in the putative "produce component" (diluted by meat and dairy products) making vegetarians subject to exposure to a unique profile of congeners.

In the UK, where most of the herbage accumulation models for PCBs are being developed, ambient contamination is lower and the projected intake from vegetables does not include congeners such as CBs 16, 18, 101, and 110 (Duarte-Davidson and Jones, 1994) which are found at relatively high levels in air and produce from other areas as well as in total diets (**Table 8.2**). Part of this may be due to low ambient levels and part due to use of historical vegetable assays which may not have included the more complete profiles associated with most of K. C. Jones' studies. Common herbage CBs 28, 44 and 52 were found in vegetables; root crops were listed as sources of many of the other congeners (Duarte-Davidson and Jones, 1994). Thus, combinations of multi-congener reports must be used to develop a complete concept of the CBs characteristic of different exposure sources. The use of a single source in **Table 8.2** (Cullen et al., 1996) is justified on the basis of a broad array of produce types and consistency with the diet surveys as well as other herbage profiles (**Appendix Table 7**).

8.5.2. Classes of PCB congeners based on human exposure

The *ortho*-chlorine-enriched CBs are more numerous, are present in much greater concentrations in human and environmental reservoirs, have multiple interrelated mechanisms of action for a broad spectrum of effects, and exhibit a broader range of physical-chemical properties. The labile congeners also tend to be *para*-chlorine-poor, making them more volatile and more mobile in the environment.

An initial effort was made to classify commonly detected (in all matrices) CBs into categories based on human exposure potential (Hansen, 1998a,b). In addition to coplanar CBs, 2 and then 3 and then 4 categories were explored. The initial categories were based on an "average fish" determined from more limited data than in **Appendix Table 11**. These categories have been modified based on more recent considerations such as **Tables 8.1, 8.2, 8.3** and **Appendix Tables 11, 12, 13.** The initial categories and candidates, which some may find preferable to the revised categories, are summarized in **Table 8.4**.

Table 8.4. Initial PCB exposure categories and CBs assigned. Coplanar CBs 77, 81, 126 and 169 are an additional category.

MAJOR CBs [25% average fish; 60-95% average human]
28, 74, 99, 105(132), 118, 128, 138(163), 153, 170, 180, 187, 194
PULSE 1 CBs [27% average fish]
8, 18, 33, 49, 52, 84, 87, 95, 101, 110, 149, 174
PULSE 2 CBs [33% average fish]
15, 40, 47+48, 56+60, 70, 82, 89, 97, 114, 129, 137, 141, 146, 177, 179, 199("201"), 203

The mammalian-persistent "Major" CBs have been variously presented as such throughout the text. The mammalian-labile "Other" congeners were divided into pulses that might be typical of fish, produce and/or airborne exposures. These made a total contribution about equal to or greater than that of the "Major" congeners in fish (27%). They also occurred at sporadic frequencies in human breast milk. Those initially perceived as most commonly encountered were termed "Pulse 1" congeners.

An additional group of about 19 constitute a large proportion of fish residues and included some pulsatile CBs from human milk profiles; these were referred to as "Pulse 2" CBs (33% of average fish) (**Table 8.4**). Pulse 1 congeners are more often selected analytes and their contributions are more reliable. Certain CBs, including some in Pulse 2, are *reported* less frequently so that actual contributions may be somewhat distorted. These are now moved to the category of "Other".

A major hazard involved in presenting lists or categories is that they may be used to simplify (abridge) complex data which include additional overlooked analytes. Nevertheless, there are so many "lists" in this work, that every CB must have been included at least twice. In order to assist in developing priorities and relationships, a *preliminary* re-categorization of PCB congeners is presented in **Figure 8.2**. The format is such that adjustments should be relatively simple. Some CBs have not yet found a permanent category and may, in fact, be appropriately included more than once. CB 118 is an important mono-*ortho*, but also a Major CB. CB 132 (when resolved from CB 105) is a Major CB, but the subsequent model presents mainly pulsatile CBs as yielding significant proportions of MeSO metabolites. CB 101 is purposely at the top of Pulse 1 because of its importance in the profiles of fish from contaminated sources; at high burdens, it is at least as persistent in fish as CB 153 (**Table 7.9**; Coristine et al., 1996). CB 77 is coplanar, but parent and/or metabolites have AhR-independent actions (Hansen, 1998; Nesaretnam et al., 1996). CB 126 also has AhR-independent actions at low doses so these may be more significant at environmental levels than AhR-dependent actions (Hansen, 1998; Li et al., 1998; Seegal and Brosch, 1999).

8.6. Model of infant exposure to PCBs in human breast milk

Infant exposure is highly dependent on maternal intake and toxicokinetics within the mother. These factors have been generically modeled for organohalogens and especially dioxin-like compounds. However, a more detailed approach needs to be used to determine the infant exposure to less-studied PCB congeners which have a high potential for neurochemical and endocrine toxicities.

As with many other media, the total PCB burden in human breast milk declined significantly between the early 1970s and the mid 1980s. Most of the attention has recently been directed toward TCDD EQs which have also been declining. Contributions to toxicity are reasonably well defined and the current lifetime accumulated daily dose (LADD) for TCDD EQs from breast milk exposure appears to be unremarkable (e.g., Norén and Meironyté, 1998; Patandin et al., 1998). The toxicological significance of the other PCBs is unknown because of the uncertainty in *actual* congeners present and a general lack of information regarding the actions of these congeners.

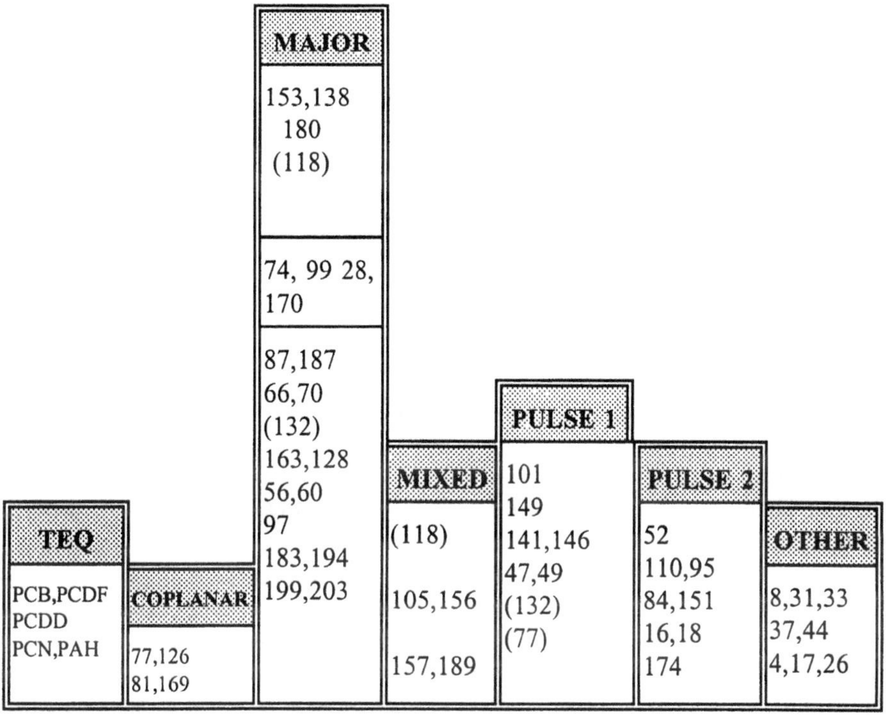

Figure 8.2. Proposed basic schematic representation for determining the relative contributions of PCB congeners to body burdens. Height is a crude representation of relative contributions. Congener assignments to categories are preliminary.

8.6.1. Transfer of PCBs through Milk: Assumptions

The most elusive variable(s) is(are) the type(s) (and proportions) of PCB congeners transferred. This will depend on sources as well as internal variables within the mother. The maternal exposure specifics can be implied from the profiles in the breast milk. The starting point, then, is to determine the profile of toxicants contained in the breast milk. Once an accurate profile is determined, basic toxicity studies can be conducted on representative CBs to fill in the void of non-dioxin-like effects so that some inference can be made regarding the toxicological significance of the exposure.

The basic model for which specific PCB congeners must be inserted and various constants must be developed is:

SOURCE → MOTHER → INFANT

The major transfer between the MOTHER and INFANT will be via breast milk; therefore, the first step is to define the amount and types of PCBs in the breast milk. Amount will eventually be determined by:

$$C_{mlk} \text{ X } V_{mlk} = \text{Amount}_{inf}$$

where C is concentration in the milk and V is the volume of milk ingested daily.

The critical issue of volume ingested per unit mass and impact on lifetime accumulated dose (LADD) must be addressed. The most critical and vulnerable developmental events occur *in utero* and during the first 6-9 months of life for a human. Therefore, this period should be considered separately from LADD calculations. USEPA (1998) recommends using an average breast milk intake of 730 ml/day (upper percentile = 1029 ml/day) for infants 1-6 months of age and this is considered rather constant up to 9 months of age. A time-weighted average of 678 ml/day is recommended for the first year since many feedings are supplemented after 5-6 months (USEPA 1998).

The average daily dose (ADD) can be determined using a 12-month (or some variation) time-span to determine contribution to LADD. This generally works out to a small proportion (4 to 12%) of the LADD for a 70 year life-span. This procedure should be adequate for most risk assessments, except for infant exposure which is, perhaps, the most critical one. Recall that swine exposed to 15 mg/kg PCB at weaning differed from controls after several months as much or more than those exposed continuously via the diet to higher LADDs (**Section 8.3.1**).

Thus, the 6-month average should be considered separately as an "Infant Exposure Norm" since this is a highly vulnerable period. Chemical insults during this period can result in toxic actions, such as interrupted development, to which more mature individuals would not be vulnerable. The manifestations may not be apparent until later in life and may be assigned a different causality. Certainly, the LADD contribution of breast-feeding is hardly significant *in the adult* for most chemicals. Such rationalized denials of significance are common because of the benefits of breast-feeding and the desire not to discourage such beneficial actions with overly conservative default assumptions. On the other hand, the 6 or 6-to-9 or 12 month ADD should be considered as a separate exposure period during a higher intake-per-weight period of greater vulnerability.

Body weights are increasing rapidly during this period, but, if a median weight of 9 kg is assumed, breast milk consumption of 730 ml in the 6-11 month period is about 80 g/kg/day (0.73 kg/9 kg). If a "70 kg man" were to consume the equivalent, he would have to eat and drink 5.6 kg (over 12 pounds) each day and would probably also experience rapid weight gain. The adult, of course, would be exposed to a variety of food types whereas the infant is exposed to the same mixture and at higher daily mass-relative doses.

The volume of breast milk during lactational exposure, then, is 180 days x 730 ml or 360 days x 678 ml and should be considered as a separate exposure event as well as being included within the LADD.

8.6.2. Infant PCB exposure through breast milk

The general topic of breast milk POPs has been studied and discussed extensively. This section will address breast milk exposure to specific PCB congeners to introduce methods that may encourage/facilitate more detailed considerations and may be useful in modeling the specific congener characteristics of other organisms and exposures. Although all of the data reviewed need to be more closely examined, a few generalizations have emerged.

It is necessary for a collective (rather than individual) effort to develop final recommendations. This single-minded effort would definitely be influenced by individual biases, oversights, and the limited perspectives of a single scientist. Nevertheless, a starting point is presented which can be readily modified by a group effort. More importantly, this highlights areas of toxicological ignorance which must be corrected before final risk assessments can be made.

Breast milk PCB residues can be more specifically characterized on the basis of "Major" and "Pulse" congeners suggested in **Figure 8.2** inserted into the simple exposure model. For the CBs to reach the milk, it will be assumed that they were present in the nursing mother (i.e., inadvertent contamination is a separate issue). The sources of the maternal burden can, for simplicity, be derived from those in **Table 8.2** plus human-like profiles in other foods. The proportion ranges of congener types in breast milk would approximate those to which the mother was exposed; however, the mixture would be somewhat modified by maternal toxicokinetics. The toxicokinetic parameters are highly congener-specific, so this requires some simple further compartmentalization within the mother.

One reason pulsatile CBs are only sporadically detected is because they tend to be more readily metabolized by mammals than by the food source in which they accumulated. The liver can be used as the representative compartment active in PCB biotransformation. Adipose tissue can be considered a reservoir receiving and releasing PCBs and MeSO metabolites depending on relative gradients with the central compartment.

This basic model is represented schematically in **Figure 8.3**. Only demonstrated major processes are considered in an attempt to preserve simplicity. The mono-*ortho* CB category can be considered to be involved in several pathways (major > coplanar > pulse 1).

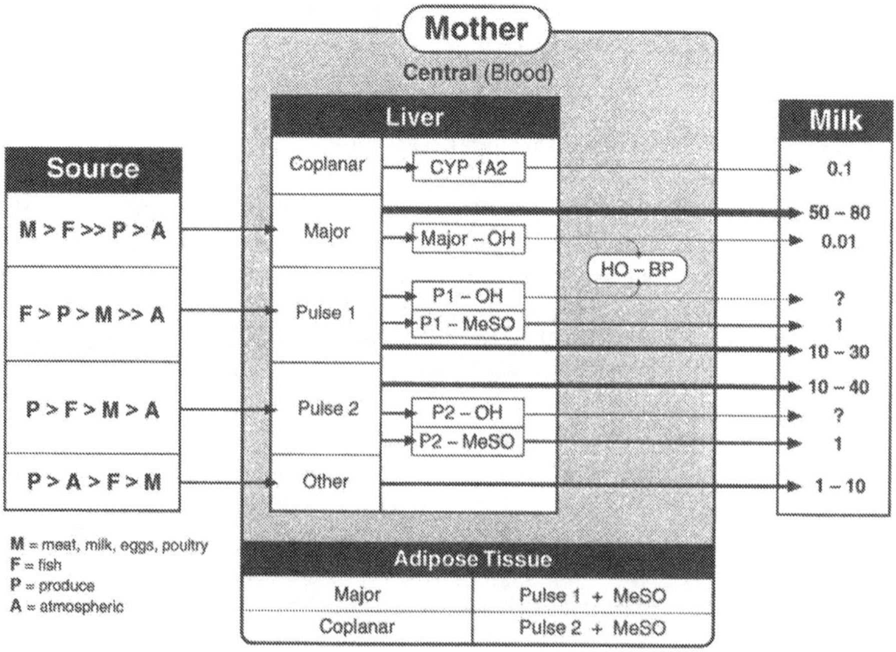

Figure 8.3. Proposed model for predicting breast milk content of major PCB congener groups (**Figure 8.2**) based on exposure sources. May also be useful in determining source(s) of sporadic deviations of breast milk PCB profiles from recognized patterns. The proportion ranges in breast milk have been derived from cited literature and data tables and the source sequences are based on profiles in **Table 8.2.**

Within the liver, coplanar and mono-*ortho* CBs induce CYP 1A proteins and are significantly and selectively sequestered by CYP 1A2 (Kuroki et al., 1986; Voorman and Aust, 1989; DeJong et al., 1993; 1995; Li et al., 1998). The coplanar CBs, especially CB 77, are also metabolized, but this route was not included in the model. This group makes a small contribution to the total profile in most breast milk, partly due to limited occurrence, induced metabolism, and liver sequestration.

The major (and mono-*ortho*) CBs are metabolized to varying degrees, but generally to a lesser extent than the pulsatile CBs. More rapid metabolism, of course, is the main basis for the lower relative proportions of pulsatile CBs. Most of the hydroxy metabolites which persist in human serum are products from the "Major" CB group and they have a decided tendency to bind to plasma proteins (**Figure 8.3**).

Hydroxylated CBs as well as glucuronides and sulfates are more hydrophilic than the parent compounds and MeSO metabolites; therefore, they are less likely to transfer to breast milk. Newsome and Davies (1996) reported an average concentration of 1.9 pg/g 4-hydroxy CB 187 as the only consistently detectable hydroxy CB in 12 breast milk samples. Because of increased polarity, however, these metabolites may be partially excluded during the extraction and clean-up procedures which are directed toward recoveries of lipophilic POPs. A breast milk model should probably treat hydroxylation as mainly a parent-PCB-interception process.

Methyl sulfonyl metabolites, on the other hand, are persistent and lipophilic and occur consistently in human milk (Newsome and Davies, 1996; Norén et al., 1996). More commonly detected products are presented in **Table 6.2**. Greater proportions of MeSO metabolites of di-, tri- and tetra-CBs have been reported in Japanese breast milk compared to the dominance of tetra-, penta- and hexa-CB MeSOs in **Table 6.2** (Haraguchi et al., 1989a; 1989b). The pulse 1 and pulse 2 congeners are generally more susceptible to the oxidative routes eventually leading to MeSOs, but smaller amounts of coplanar, mono-*ortho*, and major CBs are also converted. The MeSOs also persist in the adipose tissue.

Collective observations and data (e.g. Appendices) have been used to estimated the ranges of congener types and metabolites presented as characteristic of human breast milk (**Figure 8.3**). Exposure sources (**Table 8.2**) as modified by maternal factors can explain these profiles and predict sporadic changes.

8.7 Final considerations

Similar relationships can be readily derived for multiple pathway accumulations of specific POPs in various organisms. The focus of this monograph has been on *ortho*-rich PCBs and the author has tried to build a case for more detailed considerations of the unique individual congeners. The effort may have seemed excessive. Still, even though PCBs can be described in general terms as chlorinated aromatics with properties consistent with the generic concept of CAs, disposition and toxicity are highly congener-specific.

As should have been expected, the CBs present in breast milk are not directly proportional to those released into the environment or even to those present in maternal exposures. Proportions in maternal sources are also congener-specific. The 209 individual CBs vary in properties in a systematic, but not necessarily linear-with-IUPAC No., fashion (**Figure 2.1**). There are too many biotic factors intervening to develop direct relationships between breast milk PCB profiles and any physical-chemical constants. With limited numbers of congeners, there have been crude correlations between partition coefficients and residues in breast milk as well as various other matrices; however, the more complete CB profiles developed include congeners which deviate from the general linear trends.

One goal should be to determine if PCBs, at current levels and proportions, present a significant risk to humans and wildlife. Most correlations with adverse health effects have been ambiguous. If the subset of the population exposed to a different profile of congeners were analyzed separately as well as collectively, more direct relationships might well be revealed. Certainly, it is desirable to test outcomes against TCDD EQs. The lack of plausible relationships in other comparisons may arise from the lumping of PCBs into dioxin-like and not dioxin-like categories. It would also be relatively difficult to detect a biologically plausible relationship to specific types of actions if: 1) the profile of compounds present was not accurately represented, 2) only a specific fraction of the considered population were exposed to a subset of chemicals, and 3) that fraction of the population was grouped by factors which do not recognize the unique subset.

Effective medical intervention (e.g. in cases of high breast milk burdens) is also highly dependent on an accurate residue profile. Vitamin A supplementation presents a potential fetotoxicity, but may be indicated in cases of high TCDD EQs. Vitamin K supplementation may be indicated by higher PB EQs. Responses to pulses of readily metabolized CBs, especially in concert with CYP-inducing POPs, may be dominated by oxidative stress (e.g., Amaro et al., 1996; McLean et al., 1996). Antioxidants such as vitamin E have been shown to offer some protection to, e.g., endothelial cells (B Hennig and LW Robertson, personal communication, 1999).

The plausible associations of adverse developmental outcomes with pulsatile exposures to endocrine active PCBs and/or metabolites may prove to be insignificant at current ambient levels. It is important to develop a reasonable certainty in either case so that undue risks do not go unattenuated or that resources are not squandered on nonproblems. Hopefully, this work will encourage a more open-ended evaluation of information which may be complex only because it fails to meet the expectations of dogma developed from abridged data.

REFERENCES

Abramowitz DA. Aerobic and anaerobic biodegradation of PCBs: a review. *In* Steward GG, Russell I (eds) CRC Critical Reviews in Biotechnology, Vol 10:241-251, CRC Press, Boca Raton (1990).

Abramowitz DA, Brennan MJ, Van Dort HM, Gallagher EL. Factors influencing the rate of PCB dechlorination in Hudson River sediments. Environ Sci Technol 27:1125-1131 (1993).

Adams DD, Fendinger NJ, Glotfelty DE. Biogenic gas production and mobilization of in-place sediment contaminants by gas ebullition. *In* Baudo R, Giesy J, Muntau H (eds.) *Sediments: Chemistry and Toxicity of In-Place Pollutants*, Chelsea, MI, Lewis Publishers, Inc, pp 215-236 (1990).

Ahlborg UG, Becking GC, Birnbaum LS, Brouwer A, Derks HJGM, Feeley M, Golor G, Hanberg A, Larsen JC, Liem AKD, Safe SH, Schlatter C, Woern F, Younes M, Yrjanheikki E. Toxic equivalency factors for dioxin-like PCBs. Chemosphere 28:1049-1067 (1994).

Albro PW, Corbett JT, Schroeder JL. Quantitative characterization of polychlorinated biphenyl mixtures (Aroclors 1248, 1254 and 1260) by gas chromatography using capillary columns. J Chromatogr 205:103-111 (1981).

Albro PW, Haseman JK, Clemmer TA, Corbett BJ. Identification of the individual polychlorinated biphenyls in a mixture by gas-liquid chromatography. J Chromatogt 136:147-153 (1977). Errata in J Chromatogr 139:404 (1977).

Albro PW, Parker CE. Comparison of the compositions of Aroclor 1242 and Aroclor 1016. J Chromatogr 169:161-166 (1979).

Alvares AP, Bickers DR, Kappas A. Polychlorinated biphenyls: A new type of inducer of cytochrome P-448 in the liver. Proc Natl Acad Sci USA 70:1321-1325 (1973).

Alvares AP, Fischbein A, Anderson KE, Kappas A. Alterations in drug metabolism in workers exposed to polychlorinated biphenyls. Clin Pharmacol Therap 22:140-146 (1977).

Amaro AR, Oakley GG, Bauer U, Spielmann HP, Robertson LW. Metabolic activation of PCBs to quinones: Reactivity toward nitrogen and sulfur nucleophiles and influence of superoxide dismutase. Chem Res Toxicol 9(3):623-629 (1996).

Anderson HA, Falk C, Hanrahan L, Olson J, Burse VW, Needham L, Paschal D, Patterson D Jr, Hill RH Jr. Great Lakes Consortium. Profiles of Great Lakes Critical Pollutants: A sentinel analysis of human blood and urine. Environ Health Perspec 106:279-289 (1998).

Anderson LM, Fox SD, Riggs CW, Issaq HJ. Selective retention of polychlorinated biphenyl congeners in lung and liver after single-dose exposure of infant mice to Aroclor 1254. J Environ Pathol Toxicol Oncol 12:3-16 (1993).

Ankley GT, Cook PM, Carlson AR, Call DJ, Swenson JA, Corcoran HF, Hoke RA. Bioaccumulation of PCBs from sediments by oligochaetes and fishes: comparison of laboratory and field studies. Can J Fish Aquat Sci 49:2080-2085 (1992).

Archibeque-Engle S, Tessari JD, Winn DT. Quality assurance/quality control procedures for chlorinated hydrocarbons in human breast adipose tissue. J Toxicol Environ Health 49:589-598 (1996).

Arena SM, Halbrook RS, Arenal CA. Predicting starling chick carcass PCB concentrations from PCB concentrations in ingested animal matter. SETAC abstract (1998); Environ Toxicol Chem, submitted (1999).

Arenal CA. European starling (*Sturnus vulgaris*): Avian model and monitor of contaminant and remedial effects at Crab Orchard National Wildlife Refuge. MS Thesis, Southern Illinois University, Carbondale, IL. 123 pp. (1997).

Arenal CA, Halbrook RS. PCB and heavy metal contamination and effects in European starlings (*Sturnus vulgaris*) at a Superfund site. Bull Environ Contam Toxicol 58:254-262 (1997).

Ariyoshi N, Oguri K, Koga N, Yoshimura H, Funae Y. Metabolism of highly persistent PCB congener, 2,4,5,2',4',5'-hexachlorobiphenyl, by human CYP 2B6. Biochem and Biophys Res Commun 212(2):455-460 (1995).

Armour JA, Burke JA. Behavior of chlorinated naphthalenes in analytical methods for organochlorine pesticides and PCBs. J Assoc Offic Analyt Chem 54:175-177 (1971).

ATSDR. Toxicological Profile for Chlorodibenzofurans. US Dept Health and Human Services, Atlanta (1994).

ATSDR. Toxicological Profile for Polychlorinated Biphenyls (update). Atlanta: US Department of Health and Human Services, 1997.

Ax RL, Hansen LG. Effects of purified polychlorinated biphenyl analogs on chicken reproduction. Poult Science 54:895-900 (1975).

Bacci E, Cerejeira MJ, Gaggi C et al. Chlorinated dioxins: Volatilization from soils and bioconcentration in plant leaves. Bull Environ Contam Toxicol 48:401-408 (1992).

Bacci E, Calamari D, Gaggi C, Fanelli R, Focardi S, Morosini M. Chlorinated hydrocarbons in lichen and moss samples from the Antarctic peninsula. Chemosphere 15:747-754 (1986).

Bakke JE, Bergman Å, Larsen GL. Metabolism of 2,4',5-trichlorobiphenyl in the mercapturic acid pathway. Science 217:645-647 (1982).

Bakke JE, Feil VJ, Price CE. Replacement of chlorine with a methylsulfonyl group in the metabolism of propachlor (2-chloro-*N*-isopropylacetanilide). Biomed Mass Spectrom 3:226-229 (1976).

Ballschmiter K, Bacher R, Mennel A, Fischer R, Riehle U, Swerev M. The determination of chlorinated biphenyls, chlorinated dibenzodioxins, and chlorinated dibenzofurans by GC-MS. J High Resolut Chromatogr 15:260-270 (1992).

Ballschmiter K, Njemczyk R, Schafer W, Zoller W. Isomer-specific identification of polychlorinated benzenes (PCBz) and -biphenyls (PCB) in effluents of municipal waste incineration. Fresenius J Anal Chem 328:583-587 (1987).

Ballschmiter K, Schafer W, Buchert H. Isomer-specific identification of PCB congeners in technical mixtures and environmental samples by HRGC-ECD and HRGC-MSD, Fresenius J Anal Chem 326:253-257 (1987).

Ballschmiter K, Zell M. Analysis of polychlorinated biphenyl (PCB) by glass capillary gas chromatography: Composition of technical Aroclor and Clophen-PCB mixtures. Fresenius Z Anal Chem 302:20-31 (1980).

Barnes D, Alford-Stevens A, Birnbaum L, Kutz FW, Wood W, Patton D. Toxicity equivalency factors for PCBs? Qual Assur Gd Pract Regul Law 1:70-81 (1991).

Bedard DL, May RJ. Characterization of the polychlorinated biphenyls in the sediments of Woods Pond: Evidence for microbial dechlorination of Aroclor 1260 in situ. Environ Sci Technol 30:237-241 (1996).

Beebe L, Fox SD, Riggs CW, Park SS, Gelboin HV, Issaq HJ, Anderson LM. Persistent effects of a single dose of Aroclor 1254 on cytochromes P450IA1 and IIB1 in mouse lung. Toxicol Appl Pharmacol 114:16-24 (1992).

Benthe HF, Knop J, Schmoldt A. Aufnahme und Verteilung nach Inhalation polychlorierter Biphenyle (PCB). Toxikol 29:85-95 (1972).

Berenbaum MR, Favret C, Schuler MA. On defining "key innovations" in an adaptive radiation: Cyctochrome P450s and Papilionidae. Amer Naturalist 148(Suppl):S139-S155 (1996).

Bergen BJ, Nelson WG, Pruell RJ. Bioaccumulation of PCB congeners by blue mussels (*Mytilus edulis*) depolyed in New Bedford Harbor, Massachusetts. Environ Toxicol Chem 12:1671-1681 (1993).

Bergman Å, Larsen GL, Bakke JE. Biliary secretion, retention and excretion of five ^{14}C-labelled polychlorinated biphenyls in the rat. Chemosphere 11:149-253 (1982).

Bergman Å, Norstrom RJ, Haraguchi K, Kuroki H, Béland P. PCB and DDE methylsulfones in mammals from Canada and Sweden. Environ Toxicol Chem 13:121-128 (1994).

Bergman Å, Ellerichmann T, Franke S, Hühnerfuss H, Jakobsson E, König WA, Larsson C. Gas chromatographic enantiomer separations of chiral PCB methyl sulfones and identification of selectively retained enantiomers in human liver. Organohal Compounds 35:339-342 (1998).

Bergman Å, Athanasiadou M, Bergek S, Haraguchi K, Jensen S, Klasson-Wehler E. PCB and PCB methyl sulfones in mink treated with PCB and various PCB fractions. Ambio 21:570-576 (1992).

Bergman Å, Klasson-Wehler E, Kuroki H. Selective retention of hydroxylated PCB metabolites in blood. Environ Health Perspect 102:464-469 (1994).

Berlin M, Gage J, Holm S. Distribution and metabolism of 2,4,5,2',5'-pentachlorobiphenyl. Arch Environ Health 30:141-147 (1975).

Bertuzzi PF, Kamps L, Miles CI, Burke JA. Extraction of chlorinated pesticide residues from nonfatty samples of low moisture content. J Assoc Offic Analyt Chem 50:623-627 (1967).

Bidleman TF, Olney CE. Chlorinated hydrocarbons in the Sargasso Sea atmosphere and surface water. Science 183:516-518 (1974).

Billings RE, McMahon RE. Microsomal biphenyl hydroxylation: The formation of 3-hydroxybiphenyl and biphenyl catechol. Mol Pharmacol 14:145-154 (1978).

Birnbaum L. Distribution and excretion of 2,3,6,2',3',6'- and 2,4,5,2',4',5'-hexachlorobiphenyl in senescent rats. Toxicol Appl Pharmacol 70:262-272 (1983).

Blais JM, Froese KL, Schindler DW, Muir DCG. Assessment of PCBs in snow and lake sediments following a major release from the Alberta special waste treatment centre near Swan Hills, Alberta, Canada. Organohal Compounds 39:189-192 (1998).

Bleavins MR, Aulerich RJ, Ringer RK. Polychlorinated biphenyls (Aroclors 1016 and 1242): Effects on survival and reproduction in mink and ferrets. Arch Environ Contam Toxicol 9:627-635 (1980).

Block WD, Cornish HH. Metabolism of biphenyl and 4-chlorobiphenyl in the rabbit. J Biol Chem 234:3301-3302 (1959).

Boon JP, Duinker JC. Monitoring of cyclic organochlorines in the marine environment. Environ Monit Assess 7:189-208 (1986).

Boon J, Oostingh I, van der Meer J, Hillebrand MT. A model for the bioaccumulation of chlorobiphenyl congeners in marine mammals. Eur J Pharmacol 270:237-251 (1994).

Boon JP, Oudejans RCHM, Duinker JC. Kinetics of individual polychlorinated biphenyl (PCB) components in juvenile sole (*Solea solea*) in relation to their concentrations in food and to lipid metabolism. Comp Biochem Physiol 79C:131-142 (1984).

Boon JP, van der Meer J, Allchin CR, Law RJ, Klungsoyr J, Leonards PE, Spliid H, Storr-Hansen E, Mckenzie C, Wells DE. Concentration-dependent changes of PCB patterns in fish-eating mammals: structural evidence for induction of cytochrome P450. Arch Environ Contam Toxicol 33(3):298-311 (1997).

Borchard RE, Hansen LG, Huber WG, Metcalf RL, Welborn ME. Pharmacokinetics of Aroclor 1254 components after intravenous administration to swine and sheep. Arch Environ Contam Toxicol 2:179-192 (1974).

Borchard RE, Welborn ME, Hansen LG, Link RP, Teske RH. Apparent pharmacokinetics of PCB components in growing pigs and lambs when fed a ration containing Aroclor 1254. Arch Environ Contam Toxicol 4:226-245 (1976).

Borchard RE, Welborn ME, Wiekhorst WB, Wilson DW, Hansen LG. Pharmacokinetics of Aroclor 1254 components in swine and sheep after a single oral dose. J Pharmaceut Sci 64:1294-1302 (1975).

Borlakoglu JT, Wilkins JPG. Metabolism of di-, tri-, tetra-, penta- and hexachlorobiphenyls by hepatic microsomes isolated from control animals and animals treated with Aroclor 1254, a commercial mixture of polychlorinated biphenyls (PCBs). Comp Biochem Physiol 105C:95-106 (1993).

Bouwman CA, Seinen W, Koppe JG, van den Berg M. Effects of 2,3,7,8-tetrachlorodibenzo-*p*-dioxin or 2,2',4,4',5,5'-hexachlorobiphenyl on vitamin K-dependent blood coagulation in female germfree WAG/Rij rats. Toxicol 75:109-120 (1992).

Brandt I. Tissue localization of polychlorinated biphenyls: Chemical structures related to pattern of distribution. Acta Pharmacol Toxicol 40(Suppl II):5-108 (1977).

Brandt I, Bergman Å. PCB methyl sulphones and related compounds: Identification of target cells and tissues in different species. Chemosphere 16:1671-1676 (1987).

Brandt I, Bergman Å, Klasson Wehler E, Poellinger L, Gustafsson J-Å. Target cells for the PCB metabolite 4,4'-bis(methylsulfphonyl)-2,2',5,5'-tetrachlorobiphenyl in lung and kidney. Drug Metab Dispos 13:490-496 (1985).

Brandt I, Mohammed A, Slanina P. Persistence of 2,3,6-substituted pentachlorobiphenyls in the lung parenchyma: a new structure-dependent tissue localization of PCBs in mice. Toxicol 21:317-322 (1981).

Bright DA, Dushenko WT, Grundy LS, Reimer KJ. Evidence for short range transport of polychlorinated biphenyls in the Canadian Arctic using congener signatures of PCBs in soils. Sci Total Environ 160/161:251-263 (1995).

Brouwer A, van den Berg K. Binding of a metabolite of 3,3',4,4'-tetrachlorobiphenyl to transthyretin reduces serum vitamin A transport by inhibiting the formation of the protein complex carrying both retinol and thyroxin. Toxicol Appl Pharmacol 85:301-312 (1986).

Brown AP, Ganey PE. Neutrophil degranulation and superoxide production induced by polychlorinated biphenyls are calcium dependent. Toxicol Appl Pharmacol 131:198-205 (1995).

Brown, JF Jr. Determination of PCB metabolic, excretion and accumulation rates for use as indicators of biological responses and relative risk. Environ Sci Technol 28:2295-2305 (1994).

Brown JF Jr. Metabolic alterations of PCB residues in aquatic fauna: Distributions of cytochrome P4501a- and P4502B-like activities. Mar Environ Res 34:261-266 (1992).

Brown JF Jr, Bedard DL, Brennan MJ, Carnahan JC, Feng H, Wagner RE. PCB dechlorination in aquatic sediments. Science 236:709-712 (1987).

Brown JF Jr, Lawton RW, Morgan CB. PCB metabolism, persistence and health effects after occupational exposure: Implications for risk assessment. Chemosphere 29:2287-2294 (1994).

Brown JF Jr, Lawton RW, Ross MR, Feingold J, Wagner RE, Hamilton SB. Persistence of PCB congeners in capacitor workers and yusho patients. Chemosphere 19:829-834 (1989).

Brown JF Jr, Wagner RE. PCB movement, dechlorination, and detoxication in the Acushnet Estuary. Environ Toxicol Chem 9:1215-1233 (1990).

Bruhn R, Kannan N, Petrick G, Schulz-Bull DE, Duinker JC. CB pattern in the harbour porpoise: Bioaccumulation, metabolism and evidence for cytochrome P450 IIB activity. Chemosphere 31:3721-3732 (1995).

Bühler F, Schmid P, Schlatter C. Kinetics of PCB elimination in man. Chemosphere 17:1717-1726 (1988).

Bui QD, Weisz J. Monooxygenase mediating catechol estrogen formation by rat anterior pituitary is an estrogen-4-hydroxylase. Endocrinol 124:1085-1087 (1989).

Burgess RM, McKinney RA, Brown WA. Enrichment of marine sediment colloids with polychlorinated biphenyls: Trends resulting from PCB solubility and chlorination. Environ Sci Technol 30:2556-2566 (1996).

Burke MD, Thompson S, Elcombe CR, Halpert J, Haaparanta T, Mayer RT. Ethoxy-, pentoxy-, and benzyloxyphenoxazones and homologues: A series of substrates to distinguish between different induced cytochrome(s) P450. Biochem Pharmacol 34:3337-3345 (1985).

Burse VW, Groce DF, Caudill SP, Korver MP, Phillips DL, McClure PC, Lapeza CR Jr, Head SL, Miller DT, Buckley DJ, Nassif J, Timperi RJ, George PM. Determination of polychlorinated biphenyl levels in the serum of residents and in the homogenates of seafood from the New Bedford, Massachusetts, area: A comparison of exposure sources through pattern recognition techniques. Sci Tot Environ 144:153-177 (1994).

Busbee DL, Yoo J-SH, Norman JO, Joe CO. Polychlorinated biphenyl uptake and transport by lymph and plasma components (42073). Proc Soc Exper Biol Med 179:116-122 (1985).

Bush B, Baker FD, Tumasonis CE, Lo F-C, Houck CL. Modification of the homolog and isomer composition of a PCB mixture during passage through two biological systems. In Proceedings: Recent

Advances in the Assessment of the Health Effects of Environmental Pollution. Commission of the European Communities, Publisher, Vol. 2, pp 879-885, Luxembourg, 1975.

Bush B, Connor S, Snow J. Glass capillary gas chromatography for sensitive, accurate polychlorinated biphenyl analysis. J Assoc Offic Anal Chem 65:555-566 (1982).

Bush B, Murphy MJ, Connor S, Snow J, Barnard E. Improvements in glass capillary gas chromatographic polychlorobiphenyl analysis. J Chromatrogr Science 23:509-515 (1985a).

Bush B, Snow J, Connor S, Koblintz R. Polychlorinated biphenyl congeners (PCBs), *p,p'*-DDE and hexachlorobenzene in human milk in three areas of upstate New York. Arch Environ Contam Toxicol 14:443-450 (1985b).

Bush B, Streeter RW, Sloan RJ. PCB congeners in striped bass (*Morone saxatilis*) from marine and estuarine waters of New York State determined by capillary gas chromatography. Arch Environ Contam Toxicol 19:49-61 (1989).

Bush B, Tumasonis CF, Baker FD. Toxicity and persistence of PCB homologs and isomers in the avian system. Arch Environ Contam Toxicol 2:195-212 (1974).

Byrne JJ, Carbone JP, Hanson EA. Hypothyroidism and abnormalities in the kinetics of thyroid hormone metabolism in rats treated chronically with polychlorinated biphenyl and polybrominated biphenyl. Endocrinology 121:520-527 (1987).

Camphens J, Mackay D. Fugacity-based model of PCB bioaccumulation in complex aquatic food webs. Environ Sci Technol 31:577-583 (1997).

Canfield LM, Giuliano AR, Graver EJ. Carotenoids, retinoids, and vitamin K in human milk. In: Jensen RG (ed) *Handbook of Milk Composition*, Academic Press, San Diego, pp. 693-705 (1995).

Chang TKH, Waxman DJ. The CYP2A subfamily. *In. Cytochromes P450: Metabolic and Toxicological Aspects* (Ioannides C, ed). Boca Raton: CRC Press, 1996;99-134.

Chen PH, Luo ML, Wong CK, Chen CJ. Comparative rates of elimination of some individual polychlorinated biphenyls from the blood of PCB-poisoned patients in Taiwan. Food Chem Toxicol 20:417-425 (1982).

Chen PR, McKinney JD, Matthews HB. Metabolism of 2,4,5,2-,5'-pentachlorobiphenyl in the rat. Drug Metab Dispos 4:362-367 (1976).

Chen W, Schopflocher D, Guidotti TL, Muir DG, Lastoria C, et al. Preliminary analysis of levels of PCBs, PCDDs and PCDFs in blood in residents of Swan Hills and surrounding communities, Alberta, Canada. Organohal Compounds 39:177-180 (1998).

Chiarenzelli J, Scrudato R, Arnold G, Wunderlich M, Rafferty D. Volatilization of polychlorinated biphenyls from sediment during drying at ambient conditions. Chemosphere 33:899-911 (1996).

Chiarenzelli J, Scrudato R, Wunderlich M. Volatile loss of PCB Aroclors from subaqueous sand. Environ Sci Technol 31:597-602 (1997).

Clarkson TW. Environmental contaminants in the food chain. Am J Clin Nutr 6(suppl):682S-686S (1995).

Clevenger MA, Roberts SM, Lattin DL, Harbison RD, James RC. The pharmacokinetics of 2,2',5,5'-tetrachlorobiphenyl and 3,3',4,4'-tetrachlorobiphenyl and its relationship to toxicity. Toxicol Appl Pharmacol 100:315-327 (1989).

Cochran JW, Frame GM. Recent developments in the high resolution gas chromatography of polychlorinated biphenyls. J Chromatogr, in press (1999).

Combs GF Jr, Brewer RN, Williams NK, Richburg RW. Dieldrin dilution in visceral fat of field-contaminated broilers. Poultry Science 54:1713-1716 (1975).

Connor K, Safe S, Jefcoate CR, Larsen M. Structure-dependent induction of CYP2B by polychlorinated biphenyl congeners in female Sprague-Dawley rats. Biochem Pharmacol 50:1913-1020 (1995).

Coristine S, Haffner GD, Ciborowski JJH, Lazar R, Nanni ME, Metcalfe CD. Elimination rates of selected di-*ortho*, mono-*ortho*, and non-*ortho* substituted polychlorinated biphenyls in rainbow trout (*Oncorhynchus mykiss*). Environ Toxicol Chem 15:1382-1387 (1996).

Coyle GT, Harmon TC, Suffet IH. Aqueous solubility depression for hydrophobic organic chemicals in the presence of partially miscible organic solvents. Environ Sci Technol 31:384-389 (1997).

Cullen AC, Vorhees DJ, Altshul LM. Influence of harbour contamination on the level and composition of polychlorinated biphenyls in produce in Greater New Bedford, Massachusetts. Environ Sci Technol 30:1581-1588 (1996).

Curtis LR, Kerkvliet NI, Baecher-Steppan L, Carpenter HM. 2,3,7,8-tetrachlorodibenzo-p-dioxin pretreatment of female mice altered tissue distribution but not hepatic metabolism of a subsequent dose. Fund Appl Toxicol 14:523-531 (1990).

Dachs J, Bayona JM, Fowler SW. Flicker noise in vertical fluxes of particle-associated contaminants in the marine environment. Environ Sci Technol 30:3392-3396 (1996).

Daly JW, Jerina DM, Witkop B. Arene oxides and the NIH shift: The metabolism, toxicity and carcinogenicity of aromatic compounds. Experentia 28:1129-1164 (1972).

Darnerud PO, Brandt I, Klasson-Wehler E, Bergman Å, D'argy R, Dencker L, Sperber GO. 3,3',4,4'-tetrachloro[^{14}C]biphenyl in pregnant mice: enrichment of phenol and methyl sulphone metabolites in late gestational fetuses. Xenobiotica 16:295-306 (1986).

De Boer J, van der Valk F, Kerkhoff MAT, Hagel P, Brinkman UATh. Eight-year study on the elimination of PCBs and other organochlorine compounds from eel (*Anguilla anguilla*) under natural conditions. Environ Sci Technol 28:2242-2248 (1994).

De Jongh J, De Vito M, Nieboer R, Birnbaum L, Van den Berg M. Induction of cytochrome P450 isoenzymes after toxicokinetic interactions between 2,3,7,8-tetrachlorodibenzo-p-dioxin and 2,2',4,4',5,5'-hexachlorobiphenyl in the liver of the mouse. Fund Appl Toxicol 25:264-270 (1995).

de Jongh J, Wondergem F, Seinen W, Van den Berg M. Toxicokinetic interactions between chlorinated aromatic hydrocarbons in the liver of the C57BL/6J mouse: 1. Polychlorinated biphenyls (PCBs). Arch Toxicol 67:453-460 (1993).

den Besten C, Bennik MHJ, Bruggeman I, Schielen P, Kuper F, Brouwer A, Koeman JH, Vos JG, van Blanderen PJ. The role of oxidative metabolism in hexachlorobenzene-induced porphyria and thyroid hormone homeostasis: A comparison with pentachlorobenzene in a 13-week feeding study. Toxicol Appl Pharmacol 119:181-194 (1993).

De Voogt P, Brinkman UATh. "Production, properties and usage of polychlorinated biphenyls." In *Halogenated Biphenyls, Terphenyls, Naphthalenes, Dibenzodioxins and Related Products*, R Kimbrough, AA Jensen, eds. Amsterdam: Elsevier, 1989;3-45.

Devoto E, Fiore BJ, Millikan R, Anderson HA, Sheldon L, Sonzogni WC, Longnecker MP. Correlations among human blood levels of specific PCB congeners and implications for epidemiological studies. Am J Ind Med 32:606-613 (1997).

Dewailly E, Ryan JJ, Laliberte C, Bruneau S, Weber J-P, Gingras S, Carrier G. Exposure of remote maritime populations to coplanar PCBs. Environ Hlth Perspect 102(Suppl 1):205-209 (1994).

Dewailly E, Weber J-P, Gingras S, Laliberte C. Coplanar PCBs in human milk in the province of Quebec, Canada: Are they more toxic than dioxin for breast fed infants? Bull Environ Contam Toxicol 47:491-498 (1991).

Dewailly E, Dodin S, Verreault R, Ayotte P, Sauve L, Morin J, Brisson J. High organochlorine body burden in women with estrogen receptor-positive breast cancer. J Nation Cancer Inst 86:232-234 (1994).

DHEW Subcommittee on Health Effects of PCBs and PBBs. General summary and conclusions. Environ Health Perspec 24:173-191 (1978).

Dieter MP. Plasma enzyme activities in Coturnix quail fed graded doses of DDE, polychlorinated biphenyl, malathion and mercuric chloride. Toxicol Appl Pharmacol 27:86-98 (1974).

Ding X, Spink DC, Bhama JK, Sheng JJ, Vaz ADN, Coon MJ. Metabolic activation of 2,6-dichlorobenzonitrile, an olfactory-specific toxicant, by rat, rabbit, and human cytochromes P450. Molec Pharmacol 49:1113-1121 (1996).

Dragnev KH, Beebe LE, Jones CR, Fox SD, Thomas PE, Nims RW, Lubet RA. Subchronic dietary exposure to Aroclor 1254 in rats: Accumulation of PCBs in liver, blood and adipose tissue and its relationship to induction of various hepatic drug-metabolizing enzymes. Toxicol Appl Pharmacol 125:111-122 (1994).

Duarte-Davidson R, Burnett V, Waterhouse KS, Jones KC. A congener specific method for the analysis of polychlorinated biphenyls (PCBs) in human milk. Chemosphere 23:119-131 (1991).

Duarte-Davidson R, Jones KC. Polychlorinated biphenyls (PCBs) in the UK population: estimated intake, exposure and body burden. Sci Total Environ 151:131-152 (1994).

Duarte-Davidson R, Wilson SC, Jones KC. PCBs and other organochlorines in human tissue samples from the Welsh population: I - Adipose. Environ Pollut 84:69-77 (1994a).

Duarte-Davidson R, Wilson SC, Jones KC. PCBs and other organochlorines in human tissue samples from the Welsh population: II - Milk. Environ Pollut 84:79-87 (1994b).

Dulfer WJ, Govers HAJ, Groten JP. Kinetics and conductivity parameters of uptake and transport of polychlorinated biphenyls in the CACO-2 intestinal cell line model. Environ Toxicol Chem 17:493-501 (1998).

Dushenko WT, Grundy SL, Reimer KJ. Vascular plants as sensitive indicators of lead and PCB transport from local sources in the Canadian Arctic. Sci Tot Environ 188:29-38 (1996).

Eisenreich SJ, Baker JE, Franz T, Swanson M, Rapaport RA, Strachan WMJ, Hites RA. Atmospheric deposition of hydrophobic organic contaminants to the Laurentian Great Lakes. *In* Schnoor JL (ed.) *Fate of Pesticides and Chemicals in the Environment*, New York, John Wiley & Sons, pp. 51-78 (1992).

Eisenreich SJ, Capel PD, Robbins JA, Bourbonniere R. Accumulation and diagenesis of chlorinated hydrocarbons in Lacustrine sediments. Environ Sci Technol 23:1116-1126 (1989).

Eisenreich SJ, Looney BB, Thornton JD. Airborne organic contaminants in the Great Lakes ecosystem. Environ Sci Technol 15:30-38 (1981).

Eisler R. PCB hazards to fish, wildlife, and invertebrates: A synoptic review. Biol Rpt 85(1.7), US Fish and Wildlife Service, Laurel MD 72 pp (1986).

Erickson MD. *Analytical Chemistry of PCBs*, 2nd Edition. Boca Raton, Lewis Publishers, 1997.

Fait A, Grossman E, Self S, Jeffries J, Pellizzari ED, Emmett EA. Polychlorinated biphenyl congeners in adipose tissue lipid and serum of past and present transformer repair workers and a comparison group. Fund Appl Toxicol 12:42-55 (1989).

Falandysz J, Dembowska A, Strandberg L, Strandberg B, Bergqvist P-A, Rappe C. Congener-specific data of PCBs in some species of fish from the Gulf of Gdańsk, Baltic Sea. Organohal Compounds 32:358-363 (1997).

Falandysz J, Dembowska A, Strandberg L, Strandberg B, Bergqvist P-A, Rappe C. Non-, mono-*ortho* and total PCBs in black cormorants and their food in the Gulf of Gdańsk, Baltic Sea. Organohal Compounds 39:47-51 (1998).

Falconer RL, Bidleman TF. Vapor pressures and predicted particle/gas distributions of polychlorinated biphenyl congeners as functions of temperature and ortho-chlorine substitution. Atmosph Environ 28:547-554 (1994).

Felt GR, Mueller WF, Iatropoulos MJ, Coulston F, Korte F. Chronic toxicity of 2,5,4'-trichlorobiphenyl in young rhesus monkeys. I. Body Distribution, Elimination, and Metabolism. Toxicol Appl Pharmacol 41:619-627 (1977).

Feyereisen R, Koener JF, Farnsworth DE, Nebert DW. Isolation and sequence of cDNA encoding a cytochrome P-450 from an insecticide-resistant strain of the house fly, *Musca domestica*. Proc Natl Acad Sci USA 86:1465-1469 (1989).

Fish KM, Mayes BA, Brown JF Jr, Silkworth JB. Biochemical measurements on hepatic tissues from SD rats fed Aroclors 1016, 1242, 1254, and 1260. Toxicologist 36:87 (1997).

Focardi S, Gaggi C, Chemello G, Bacci E. Organochlorine residues in moss and lichen samples from two Antarctic areas. Polar Rec 27:241-244 (1991).

Focardi S, Leonzio C, Fossi C. Variations in polychlorinated biphenyl congener composition in eggs of Mediterranean water birds in relation to their position in the food chain. Environ Pollut 52:243-255 (1988).

Fox GA, Kennedy SW, Trudeau S, Bishop CA, Wayland M. Hepatic porphyrin patterns in birds as a promising measure of effect and bioavailability of PCBs and other HAHs in water and sediments. Organohal Compounds 33:366-370 (1997).

Fox K, Zauke G-P, Butte W. Kinetics of bioconcentration and clearance of 28 polychlorinated biphenyl congeners in Zebrafish (*Brachydanio rerio*). Ecotoxicol Environ Safety 28:99-109 (1994).

Frame GM. A collaborative study of 209 PCB congeners and 6 Aroclors on 20 different HRGC columns. 1. Retention and coelution database. Fresenius J Anal Chem 357:701-713 (1997).

Frame GM, Cochran JW, Bøwadt SS. Complete PCB congener distributions for 17 Aroclor mixtures determined by 3 HRGC systems optimized for comprehensive, quantitative, congener-specific analysis. J High Resol Chromatogr 19:657-668 (1996a).

Frame GM, Wagner RE, Carnahan JC, Brown JF, May RJ, Smullen LA, Bedard DL. Comprehensive, quantitative, congener-specific analyses of eight Aroclors and complete PCB congener assignments on DB-1 capillary GC columns. Chemosphere 33:603-623 (1996b).

Fries GF, Marrow GS Jr, Gordon CH. Long-term studies of residue retention and excretion by cows fed a polychlorinated biphenyl (Aroclor 1254). J Agr Food Chem 21(1):117-121 (1973).

Froese KL, Blais JM, Muir DCG. Conifer forest vegetation as an indicator of PCB exposure in the region of Swan Hills, Alberta, Canada. Organohal Compounds 39:185-188 (1998).

Froese KL, Verbrugge DA, Ankley GT, Niemi GJ, Larsen CP, Giesy JP. Bioaccumulation of polychlorinated biphenyls from sediments to aquatic insects and tree swallow eggs and nestlings in Saginaw Bay, Michigan, USA. Environ Toxicol Chem 17(3):484-492 (1998).

Fry DM. Reproductive effects in birds exposed to pesticides and industrial chemicals. Environ Health Perspec 103(Suppl 7):165-171 (1995).

Furst P, Furst C, Wilmers K. Human milk as a bioindicator for body burden of PCDDs, PCDFs, organochlorine pesticides and PCBs. Environ Hlth Perspect 102(Suppl 1):187-193 (1994).

Gabos S, Schopflocher D, Muir DG, Schindler D, Guidotti TL et al. Levels of PCBs, PCDDs and PCDFs in deer and moose in Alberta, Canada following accidental release from a special waste treatment center. Organohal Compounds 39:169-172 (1998).

Gage JC, Holm S. The influence of molecular structure on the retention and excretion of PCBs by the mouse. Toxicol Appl Pharmacol 36:555-560 (1976).

Gaggi C, Bacci E, Calamari D, Fanelli R. Chlorinated hydrocarbons in plant foliage: an indication of the tropospheric contamination level. Chemosphere 14:1673-1686 (1985).

Gan DR, Berthouex PM. Disappearance and crop uptake of PCBs from sludge-amended farmland. Water Environ Res 66:54-69 (1994).

Ganem LG, Jefcoate CR. Endocrine factors modulate the phenobarbital-mediated induction of cytochromes P450 and phase II enzymes in a similar strain-dependent manner. Toxicol Appl Pharmacol 150:68-75 (1998).

Gardner AM, Chen JT, Roach JA, Ragelis EP. Polychlorinated biphenyls: Hydroxylated urinary metabolites of 2,5,2',5'-tetrachlorobipenyl identified in rabbits. Biochem Biophys Res Commun 55:1377-1383 (1973).

Genoni GP. Influence of the energy relationships of organic compounds on toxicity to the cladoceran *Daphnia magna* and the fish *Pimephales promelas*. Ecotoxicol Environ Safety 36:27-37 (1997).

Gerhard I, Daniel V, Link S, Monga B, Runnebaum B. Chlorinated hydrocarbons in women with repeated miscarriages. Environ Health Persp 106:675-681 (1998).

Gerstenberger SL, Gallinat MP, Dellinger JA. Polychlorinated biphenyl congeners and selected organochlorines in Lake Superior Fish. Environ Toxicol Chem 16:2222-2228 (1997).

Gerstenberger SL, Tavris DR, Hansen LK, Pratt-Shelley J, Dellinger JA. Concentrations of blood and hair mercury and serum PCBs in an Ojibwa population that consumes Great Lakes Region fish. Clin Toxico, 35:377-386 (1997).

Geyer HJ, Rimkus GG, Scheunert I, Kaune A, Kettrup A, Zeeman M, Muir DCG, Hansen LG, Mackay D. Bioaccumulation and occurrence of endocrine-disrupting chemicals (EDCs), persistent organic pollutants (POPs), and other organic compounds in fish and other organisms including human. *In* Hutzinger O, Beek B (eds), *Handbook of Environmental Chemistry*, Vol. 2, Part J, Bioaccumulation, Springer-Verlag, 1999 (in press).

Ghiasuddin SM, Menzer RE, Nelson JO. Metabolism of 2,5,2'-trichloro, 2,5,2',5'-tetrachloro-, and 2,4,5,2',5'-pentachlorobiphenyl in rat hepatic microsomal systems. Toxicol Appl Pharmacol 36:187-194 (1976).

Gilday D, Bellwar GD, Sanderson JT, Janz DM, Rifkind AB. 2,3,7,8-tetrachorodibenzo-*p*-dioxin (TCDD) induces hepatic cytochrome P450-dependent arachidonic acid epoxygenation in diverse avian orders: Regioisomer selectivity and immunochemical comparison of the TCDD-induced P450s to CYP1A4 AND 1A5. Toxicol Appl Pharmacol 150:106-116 (1998).

Gladen BC, Rogan WJ. Effects of perinatal polychlorinated biphenyls and dichlorodiphenyl dichloroethene on later development. J Pediatr 119:58-63 (1991).

Glausch A, Hahn J, Schurig V. Enantioselective determination of chiral 2,2',3,3',4,6'-hexachlorobiphenyl (PCB 132) in human milk samples by multidimensional gas chromatography/electron capture detection and by mass spectrometry. Chemosphere 30:2079-2085 (1995).

Godard CA, Miller CA, Morre MJ, Dickerson RL, Stegeman JJ. CYP 1A1/2 and CYP2B-like activities and CYP1A immunodetection in three marine mammal species. Toxicologist 36:137 (1997).

Goldberg E. Synthetic organohalides in the sea. Proc R Soc Lond B 189:277-289 (1975).

Goldstein, JA. The structure-activity relationships of halogenated biphenyls as enzyme inducers. Ann NY Acad Sci 320:164-178 (1979).

Goto M, Hattori M, Sugiura K. Metabolism of pentachloro- and hexachloro-biphenyls in the rat. Chemosphere 3:177-180 (1975).

Grant DL, Phillips WEJ, Villeneuve DC. Metabolism of a polychlorinated biphenyl (Aroclor 1254) mixture in the rat. Bull Environ Contam Toxicol 6:102-112 (1971).

Greb W, Klein W, Coulston F, Golberg L, Korte F. *In vitro* metabolism of polychlorinated biphenyls-^{14}C. Bull Environ Contam Toxicol 13:424-432 (1975).

Green NJL, Jones KC, Harwood J. Contribution of coplanar and non-coplanar polychlorinated biphenyls to the toxic equivalence of grey seal (*Halichoerus grypus*) milk. Chemosphere 33:1273-1281 (1996).

Grote W, Schmold A, Dammann GG. The metabolism of foreign compounds in rats after treatment with polychlorinated biphenyls (PCBs). Biochem Pharmacol 24:1121-1125 (1975).

Guiney PD, Peterson RE, Melancon MJ Jr, Lech JJ. The distribution and elimination of 2,5,2',5'-[^{14}C]tetrachlorobiphenyl in rainbow trout (*Salmo gairdneri*). Toxicol Appl Pharmacol 39:329-338 (1977).

Guitart R, Puig R, Gomez-Catalan J. Requirement for a standardized nomenclature criterium for PCBs: Computer-assisted assignment of correct congener denomination and numbering. Chemosphere 27:1451-1459 (1993).

Hahn ME. The aryl hydrocarbon receptor: A comparative perspective. Comparative Biochemistry and Physiology Special Issue on Forms and Functions of Cytochrome P450. In press (1998).

Hahn ME, Lamb TM, Schultz ME, Smolowitz RM, Stegeman JJ. Cytochrome P4501A induction and inhibition by 3,3',4,4'-tetrachlorobiphenyl in an Ah receptor-containing fish hepatoma cell line (PLHC-1). Aquat Toxicol 26:185-208 (1993).

Hajjar NP, Hodgson E. Flavin adenine dinucleotide-dependent monooxygenase: Its role in the sulfoxidation of pesticides in mammals. Science 209:1134-1135 (1980).

Hakansson H. The role of retinoids in dioxin toxicity. Organohal Compounds 34:402-405 (1997).

Hansen LG. Selective accumulation and depletion of PCB components: Food animal implications. Ann NY Acad Sci 320:238-246 (1979).

Hansen LG. Food chain modification of the composition and toxicity of polychlorinated biphenyl (PCB) residues. Rev Environ Toxicol 3:149-212 (1987a).

Hansen LG. Environmental toxicology of PCBs. Environ Toxicol Series 1:15-48 (1987b).

Hansen LG. Swan Hills Process Gas Release: Environmental impact assessment. Report to Alberta Environmental Protection. Edmonton, November, 1997.

Hansen LG. Stepping backward to improve assessment of PCB congener toxicities. Environ Health Perspect 106(Suppl 1):171-189 (1998a).

Hansen LG. Superfund Nonpoint Chemical Mixtures: Risk assessment of PCB mixtures in breast milk. Report to USEPA-NCEA, Cincinnati, OH. June 2, 1998 (1998b).

Hansen LG, Dorn SB, Sundlof SM, Vogel RS. Toxicity, accumulation and depletion of hexachlorobenzene in laying chickens. J Agric Food Chem 26:1369-1374 (1978).

Hansen LG, Kapoor IP, Metcalf RL. Biochemistry of selective toxicity and biodegradability: comparative *O*-dealkylation by aquatic organisms. Comp Gen Pharmac 3:339-344 (1972).

Hansen LG, Byerly CS, Metcalf RL, Bevill RF. Effect of a PCB mixture on swine reproduction and tissue residues. Am J Vet Res 36:23-26 (1975).

Hansen LG, Green D, Cochran J, Vermette S, Bush B. Chlorobiphenyl (PCB) composition of extracts of subsurface soil, superficial dust and air from a contaminated landfill. Fres J Anal Chem 357:442-448 (1997).

Hansen LG, Li MH, Saeed A, Bush B. Environmental Polychlorinated Biphenyls: Acute toxicity of landfill soil extract to female prepubertal rats. Arch Environ Contam Toxicol 29:334-343 (1995).

Hansen LG, O'Keefe PW. Polychlorinated dibenzofurans and dibenzo-*p*-dioxins in subsurface soil, superficial dust, and air extracts from a contaminated landfill. Arch Environ Contam Toxicol 31:271-276 (1996).

Hansen LG, Storr-Hansen E. Accumulation of Polychlorinated Biphenyl (PCB) Congeners in the House Fly (Musca domestica, L.). Bull Environ Contam Toxicol 48:95-100 (1992).

Hansen LG, Sullivan JM, Neff CC, Sanders PE, Lambert RJ, Beasley VR, Storr-Hansen E. PCB contamination of domestic turkeys from building materials. J Agric Food Chem 37:135-139 (1989).

Hansen LG, Tuinstra LGMTh, Kan CA, Strik JJTWA, Koeman JH. Accumulation of chloro-biphenyls in chicken fat and liver after feeding Aroclor 1254 directly or fat from swine fed Aroclor 1254. J Agric Food Chem 31:254-260 (1983).

Hansen LG, Washko PW, Tuinstra LGMTh, Dorn SB, Hinesly TD. Polychlorinated biphenyl, pesticide, and heavy metal residues in swine foraging on sewage sludge amended soils. J Agric Food Chem 29:1012-1017 (1981).

Hansen LG, Welborn ME. Distribution, dilution and elimination of polychlorinated biphenyl analogs in growing swine. J Pharmaceut Sci 66:497-501 (1977).

Hansen LG, Welborn ME, Borchard RE, Teske RH, Metcalf RL. Tissue distribution of PCB components in swine and sheep fed three different rations containing Aroclors 1242 and 1254. Arch Environ Contam Toxicol 5:257-278 (1977).

Hansen LG, Wiekhorst WB, Simon J. Effects of dietary Aroclor 1242 on channel catfish and the selective accumulation of PCB components. J Fish Res Board Can 33:1343-1352 (1976).

Hansen LG, Wilson DW, Metcalf RL, Welborn ME. Residues of PCB components in broiler cockerels receiving two Aroclors in three dietary variations. J Agric Food Chem 24:256-261 (1976).

Haraguchi K, Kato Y, Kimura R, Masuda Y. Comparative study on formation of hydroxy and sulfur-containing metabolites from different chlorinated biphenyls with 2,5-substitution in rats. Drug Metab Dispo 25:845-852 (1997).

Haraguchi K, Kato Y, Kimura R, Masuda Y. Hydroxylation and methylthiolation of mono-ortho-substituted polychlorinated biphenyls in rats: Identification of metabolites with tissue affinity. Chem Res Toxicol 11:1508-1515 (1998).

Haraguchi K, Kuroki H, Masuda Y. Determination of hydroxy methylsulfone metabolites of PCB in Yusho patients. Chemosphere 16:2033-2038 (1987).

Haraguchi K, Kuroki H, Masuda Y. Capillary gas chromatographic analysis of methylsulphone metabolites of polychlorinated biphenyls retained in human tissues. J Chromatogr A 361:239-252 (1986).

Haraguchi K, Kuroki H, Masuda Y. Polychlorinated biphenyl methylsulfone congeners in human tissues: Identification of methylsulfonyl dichlorobiphenyls. Chemosphere 18:477-484 (1989).

Haraguchi K, Kuroki H, Masuda Y. Occurrence and distribution of chlorinated aromatic methylsulphones and sulphoxides in biological samples. Chemosphere 19:487-492 (1989).

Haraguchi K, Kuroki H, Masuda Y, Koga N, Kuroki J, Hokama Y, Yoshimura H. Toxicological evaluation of sulfur-containing metabolites of 2,2,2',5'-tetrachlorobiphenyl in rats. Chemosphere 14:1755-1762 (1985).

Hardell L, van Bavel B, Lindstrom G, Frederikson H, Hagberg H, Liljegren G, Nordstrom M, Johansson B. Higher concentrations of specific polychlorinated biphenyl congeners in adipose tissue from non-Hodgkin's lymphoma patients compared with controls without a malignant disease. Internat J Oncology 9:603-608 (1996).

Harner T, Mackay D. Measurement of Octanol-Air partition coefficients for chlorobenzenes, PCBs, and DDT. Environ Sci Technol 29:1599-1606 (1995).

Harner T, Mackay D, Jones KC. Model of the long-term exchange of PCBs between soil and the atmosphere in the Southern U.K. Environ Sci Technol 29:1200-1209 (1995).

Hattemer-Frey HA, Travis CC. Assessing the extend to human exposure through the food chain to pollutants emitted from municipal solid waste incinerators. *In* Hattemer-Frey HA, Travis CC (eds) *Health Effects of Municipal Waste Incineration*, CRC Press, Boca Raton, pp. 83-101 (1991).

Hebert CE, Weseloh DV, Kot L, Glooschenko V. Organochlorine contaminants in a terrestrial foodweb on the Niagara Peninsula, Ontario, Canada 1987-89. Arch Environ Contam Toxicol 26:356-366 (1994).

Heidmann WA, Beyerbach M. Characterization of PCB mixtures found in biological samples by three parameters derived from the fractions of 45 congeners. Chemosphere 18:2303-2309 (1989).

Heit M, Klusek C. Evidence of deposition of anthropogenic pollutants in remote Rocky Mountain lakes. Water, Air, Soil Pollut. 22:403-416 (1984).

Hermanson MH, Hites RA. Long-term measurements of atmospheric polychlorinated biphenyls in the vicinity of Superfund dumps. Environ Sci Technol 23:1253-1258 (1989).

Hess P, de Boer J, Cofino WP, Leonards PEG, Wells DG. Critical review of the analysis of non- and mono-*ortho* chlorobiphenyls. J. Chromatogr A 703:417-465 (1995).

Himbergl KK, Pakarinen P. Atmospheric PCB deposition in Finland during 1970s and 1980s on the basis of concentrations in ombrotrophic peat mosses (Sphagnum). Chemosphere 29:431-440 (1994).

Hinz R, Matsumura F. Comparative metabolism of PCB isomers by three species of fish and the rat. Bull Environ Contam Toxicol 18:631-639 (1977).

Hippelein M, McLachlan MS. Soil/Air partitioning of semivolatile organic compounds. 1. Method development and influence of physical-chemical properties. Environ Sci Techol 32:310-316 (1998).

Hodgson E. Microsomal monooxygenases. *In* Kerkut GA, Gilbert LI (eds), *Comprehensive Insect Physiology Biochemistry and Pharmacology*, pp 225-321, Pergamon, Oxford (1985).

Hornbuckle KC, Achman DR, Eisenreich SJ. Over-water and over-land polychlorinated biphenyls in Green Bay, Lake Michigan. Environ Sci Technol 27:87-98 (1993).

Hornshaw TC, Aulerich RJ, Johnson HE. Feeding Great Lakes fish to mink: Effects on mink and accumulation and elimination of PCBs by mink. J Toxicol Environ Hlth 11:933-946 (1983).

Huang S, Gibson GG. Differential induction of cytochromes P450 and cytochrome P450-dependent arachidonic acid metabolism by 3,4,5,3',4'-pentachlorobiphenyl in the rat and guinea pig. Toxicol Appl Pharmacol 108:86-95 (1991).

Huang S, Gibson GG. Species and congener specific induction of hepatic cytochrome P450 4A by polychlorinated biphenyls. Biochem Pharmacol 43:637-639 (1992).

Humphrey HEB. The human population--An ultimate receptor for aquatic contaminants. Hydrobiol 149:75-80 (1987).

Humphrey HEB, Gardiner JC, O'Keefe M, Pandya JR, Sweeney AM, Gasior DM, McCaffrey RJ, Schantz SL. PCB congener profile in the serum of humans consuming Great Lakes fish. Env Health Perspec, in press (1999).

Hung C-F, Holzmacher R, Connolly E, Berenbaum MR, Schuler MA. Conserved promotor elements in the *CYP6B* gene family suggest common ancestry for cytochrome P450 monooxygenases mediating furanocuoumarin detoxification. Proc Natl Acad Sci USA, 93:12200-12205 (1996).

Hunter DJ, Hankinson SE, Laden F, Colditz GA, Manson JE, Willett WC, Speizer FE, Wolff MS. Plasma organochlorine levels and the risk of breast cancer. N Engl J Med 337:1253-1258 (1997).

Hutzinger O, Jamieson WD, Safe S, Paulmann L, Ammon R. Identification of metabolic dechlorination of highly chlorinated biphenyl in rabbit. Nature 252:698-699 (1974).

Hutzinger O, Safe S, Zitko V. *The Chemistry of PCB's*. Cleveland: CRC Press, 1974.

Ikegwuonu FI, Ganem LG, Larsen MC, Shen X, Jefcoate CR. The regulation by gender, strain, dose and feeding status of the induction of multiple forms of cytochrome P450 isozymes in rat hepatic microsomes by 2,4,5,2',4',5'-hexachlorobiphenyl. Toxicol Appl Pharmacol 139:33-41 (1996).

Ioannides C (ed). *Cytochromes P450. Metabolic and Toxicological Aspects*. Boca Raton, CRC Press, 411 pp. (1996).

Ishida C, Koga N, Hanioka N, Saeki HK, Yoshimura H. Metabolism in vitro of 3,4,3',4'- and 2,5,2',5'-tetrachlorobiphenyl by rat liver microsomes and highly purified cytochrome P450. J Pharmacobiodyn (Japan) 14(5):276-284 (1991).

Jackson LJ, Schindler DE. Field estimates of net trophic transfer of PCBs from prey fishes to Lake Michigan salmonids. Environ Sci Technol 30:1861-1865 (1996).

Jacobson JL, Humphrey HEB, Jacobson SW, Schantz SL, Mullin MD, Welch R. Determinants of PCBs, PBBs and DDT levels in the sera of young children. Am J Public Health 79:1401-1404 (1989).

Jacobson JL, Jacobson SW. Intellectual impairment in children exposed to polychlorinated biphenyls in utero. N Eng J Med 335:783-789 (1996).

Jan J, Tratnik M, Kenda A. Atmospheric contamination with polychlorinated biphenyls in Bela Krajina (Yugoslavia); emissions from industrial plant, landfill, and river areas. Chemosphere 17:809-813 (1988).

Jansson B, Bergman Å. Sulfur-containing derivatives of hexachlorobenzene (HCB) metabolites in the rat. Chemosphere 3:257-268 (1978).

Jensen AA. Polychlorobiphenyls (PCBs), polychlorodibenzo-p-dioxins (PCDDs) and polychlorodibenzofurans (PCDFs) in human milk, blood and adipose tissue. Sci Tot Environ 64:259-293 (1987).

Jensen S, Jansson B. Anthropogenic substances in seal from the Baltic: Methyl sulfone metabolites of PCB and DDE. Ambio 5:257-260 (1976).

Jensen S, Sundström G. Metabolic hydroxylation of a chlorobiphenyl containing only isolated unsubstituted positions--2,2',4,4',5,5'-hexachlorobiphenyl. Nature 251:219-220 (1974).

Jensen S, Sundström G. Structures and levels of most chlorobiphenyls in two technical PCB products and in human adipose tissue. Ambio 3:70-76 (1974).

Johansen HR, Becher G, Polder A, Skaare JU. Congener-specific determination of polychlorinated biphenyls and organochlorine pesticides in human milk from Norwegian mothers living is Oslo. J Toxicol Environ Health 42:157-171 (1994).

Johansson M, Nilsson S, Lund B-O. Interactions between methylsulfonyl PCBs and the glucocorticoid receptor. Environ Health Persp 106:769-772 (1998).

Jones KC. Determination of polychlorinated biphenyls in human foodstuffs and tissues: Suggestions for a selective congener analytical approach. Sci Total Environ 68:141-159 (1988).

Jones KC, Duarte-Davidson R, Cawse PA. Changes in the PCB concentration of United Kingdom air between 1972 and 1992. Environ Sci Technol 29:272-275 (1995).

Jones KC, Sanders G, Wild SR, Burnett V, Johnston AE. Evidence for a decline of PCBs and PAHs in rural vegetation and air in the United Kingdom. Nature 356:137-140 (1992).

Kamei K, Matsuda M, Momose A. New sulfur containing metabolites of caffeine. Chem Pharm Bull 23:683-684 (1975).

Kaminski LS, Kennedy MW, Adams SM, Guengerich FP. Metabolism of dichlorobiphenyls by highly purified isozymes of rat liver cytochrome P-450. Biochem 20:7379-7384 (1981).

Kannan N, Reusch RBH, Schulz-Bull DE, Petrick G, Duinker JC. Chlorobiphenyls: Model compounds for metabolism in food chain organisms and their potential use as ecotoxicological stress indicators by application of the metabolic slope concept. Environ Sci Technol 29:1851-1859 (1995).

Kapahi R. Modeling the dispersion of toxic air pollutants emitted from municipal solid waste incinerators. *In*: *Health Effects of Municipal Waste Incineration*, HA Hattemer-Frey, C Travis (Ed), CRC Press, Boca Raton, FL p. 67-82 (1991).

Kato Y, Haraguchi K, Kawashima M, Yamada S, Masuda Y, Kimura R. Induction of hepatic microsomal drug-metabolizing enzymes by methylsulphonyl metabolites of polychlorinated biphenyl congeners in rats. Chemico-Biol Interact 95:257-268 (1995).

Kato Y, Haraguchi K, Tomiyasu K, Saito H, Isogai M, Masuda Y, Kimura R. Structure-dependent induction of CYP2B1/2 by 3-methylsulfonyl metabolites of polychlorinated biphenyl congeners in rats. Environ Toxicol Pharmacol 3:137-144 (1997).

Kato Y, Kenne K, Haraguchi K, Masuda Y, Kimura R, Warngard L. Inhibition of cell-cell communication by methylsulfonyl metabolites of polychlorinated biphenyl congeners in rat liver epithelial IAR 20 cells. Arch Toxicol 72:178-182 (1998).

Kihlstrom JE, Olsson M, Jensen S, Johansson A, Ahlbom J, Bergman Å. Effects of PCB and different fractions of PCB on the reproduction of the mink (*Mustela vison*). Ambio 21:563-569 (1992).

Kiviranta H, Purkunen R, Vartiainen T. Levels of PCDD/Fs and PCBs in human milk in 1994 in Finland: decrease in concentrations from 1987 to 1994. Organohalogen Compounds 38:121-124 (1998).

Klaassen CD, Rozman K. Absorption, distribution, and excretion of toxicants. *In* Amdur MO, Doull JD, Klaassen CD (eds) *Casarett and Doull's Toxicology: The Basic Science of Poisons*, Fourth Edition. New York, Pergamon Press, pp 50-87 (1991).

Klasson-Wehler E, Bergman Å, Kowalski B, Brandt I. Metabolism of 2,3,4',6-tetrachlorobiphenyl: formation and tissue localization of mercapturic acid pathway metabolites in mice. Xenobiotica 17:477-486 (1987).

Klasson Wehler E, Bergman Å, Brandt I, Darnerud PO, Wachtmeister CA. 3,3',4,4'-Tetrachlorobiphenyl: Excretion and tissue retention of hydroxylated metabolites in the mouse. Drug Metab Dispo 17:441-448 (1989).

Klasson Wehler E, Hovander L, Lund B-O. 2,2',4,5,5'-Pentachlorobiphenyl: Comparative metabolism in mink (*Mustela vison*) and mouse. Chem Res Toxicol 9:1340-1349 (1996).

Klasson Wehler E, Jönsson CJ, Bergman Å, Brandt I, Darnerud PO. 3,3',4,4'-Tetrachlorobiphenyl and 3,3',4,4',5-pentachlorobiphenyl: Tissue localization and metabolic fate in the mouse. Chemosphere 19:809-812 (1989).

Klasson Wehler E, Lindberg L, Jönsson CJ, Bergman Å. Tissue retention and metabolism of 2,3,3',4,4'-pentachlorobiphenyl in mink and mouse. Chemosphere 27:2397-2412 (1993).

Knap AH, Binkley KS. Chlorinated organic compounds in the troposphere over the western North Atlantic Ocean measured by aircraft. Atmospheric Environment 25:1507-1516 (1991).

Kodavanti PRS, Ward TR, Derr-Yellin EC, Mundy WR, Casey AC, Bush B, Tilson HA. Congener-specific distribution of polychlorinated biphenyls in brain regions, blood, liver and fat of adult rats following repeated exposure to Aroclor 1254. Toxicol Appl Pharmacol 153:199-210 (1998).

Koga N, Beppu M, Ishida C, Yoshimura H. Further studies on metabolism *in vivo* of 3,4,3',4'-tetrachlorobiphenyl in rats: identification of minor metabolites in rat faeces. Xenobiotica 19:1307-1318 (1989).

Koga N, Beppu M, Yoshimura H. Metabolism *in vivo* of 3,4,5,3',4'-pentachlorobiphenyl and toxicological assessment of the metabolite in rats. J Pharmacol Bio-Dyn 13:497-506 (1990).

Koopman-Esseboom C, Huisman M, Weisglas-Kuperus N, van der Paauw CG, Tuinstra LGMTh, Boersma ER, Sauer PJJ. PCB and dioxin levels in plasma and human milk of 418 Dutch women and their infants. Predictive value of PCB congener levels in maternal plasma for fetal and infant's exposure to PCBs and dioxins. Chemosphere 28:1721-1732 (1994).

Koopman-Esseboom C, Weisglas-Kuperus N, de Ridder MA, Van der Paauw CG, Tuinstra LG, Sauer PJ. Effects of polychlorinated biphenyl/dioxin exposure and feeding type on infants' mental and psychomotor development. Pediatr 97:700-706 (1996).

Korach KS, Sarver P, Chae K, McLachlan JA, McKinney JD. Estrogen receptor binding activity of polychlorinated hydroxybiphenyls: Conformationally restricted structural probes. Mol Pharmacol 33:120-126 (1988).

Korrick SA, Altshul L. High breast milk levels of polychlorinated biphenyls (PCBs) among four women living adjacent to a PCB-contaminated waste site. Environ Health Persp 106:513-518 (1998).

Koss G, Meyer-Rogge D, Seubert S, Seubert A, Losekam M. 2,2',3',4,4',5,5'-Heptachlorobiphenyl (PCB 180)—on its toxicokinetics, biotransformation and porphyrinogenic action in female rats. Arch Toxicol 67:651-654 (1993).

Kramer VJ, Helferich WG, Bergman Å, Klasson-Wehler E, Geisy JP. Hydroxylated polychlorinated biphenyl metabolites are anti-estrogenic in a stably transfected human breast adenocarcinoma (MCF7) cell line. Toxicol Appl Pharmacol 144:363-376 (1997).

Krauthacker B, Kralj M, Tkalčević B, Reiner E. Levels of HCH, p,p'-DDE, p,;'-DDT and PCBs in human milk from a continental town in Croatio, Yugoslavia. Int Arch Occup Environ Health 58:69-74 (1986).

Krieger N, Wolff MS, Hiatt RA, Rivera M, Vogelman J, Orentreich N. Breast cancer and serum organochlorines: a prospective study among white, black and asian women. J Natl Canc Inst 86:589-599 (1994).

Kubiak, TJ, HJ Harris, LM Smith, TR Schwartz, DR Stalling, JA Trick, L Sileo, DE Docherty, TC Erdman. Microcontaminants and reproductive impairment of the Forster's tern on Green Bay, Lake Michigan--1983. Arch Environ Contam Toxicol 18:706-727 (1989).

Kucklick JR, Bidleman TF, McConnell LL, Walla MD, Ivanov GP. Organochlorines in the water and biota of Lake Baikal, Siberia. Environ Sci Technol 28:31-37 (1994).

Kuroki H, Masuda Y. Structures and concentrations of the main components of polychlorinated biphenyls retained in patients with Yusho. Chemosphere 6:469-474 (1977).

Kuroki J, Koga N, Yoshimura H. High affinity of 2,3,4,7,8-pentachlorodibenzofuran to cytochrome P450 in the hepatic microsomes of rats. Chemosphere 15:731-738 (1986).

Kuwabara K, Yakushiji T, Watanabe I, Yoshida S, Yoyama K, Kunita N. Increase in the human blood PCB levels promptly following ingestion of fish containing PCBs. Bull Environ Contam Toxicol 21:273-278 (1979).

Kwajiri K, Hayashi S-I. The CYP 1 family. *In. Cytochromes P450: Metabolic and Toxicological Aspects* (Ioannides C, ed). Boca Raton: CRC Press, 1996;77-97.

Lake CA, Lake JL, Haebler R, McKinney R, Boothman WS, Sadove SS. Contaminant levels in harbor seals from the northeastern United States. Arch Environ Contam Toxicol 29:128-134 (1995).

Lake JL, McKinney R, Lake CA, Osterman FA, Heltshe J. Comparisons of patterns of polychlorinated biphenyl congeners in water, sediment, and indigenous organisms from New Bedford Harbor, Massachusetts. Arch Environ Contam Toxicol 29:207-220 (1995).

Lans MC, Klasson Wehler E, Willemsen M, Meussen E, Safe S, Brouwer A. Structure-dependent, competitive interaction of hydroxy-polychlorobiphenyls, -dibenzo-*p*-dioxins and -dibenzofurans with human transthyretin. Chem-Biol Interact 88:7-21 (1993).

Lans MC, Spiertz C, Brouwer A, Koeman JH. Different competition of thyroxine binding to transthyretin and thyroxine-binding globulin by hydroxy-PCBs, PCDDs, and PCDFs. European J Pharmacol [Environ Toxicol Pharmacol Section] 270:129-136 (1994).

Larsen BR. HRGC separation of PCB congeners. Critical Review. J High Resolut Chromatogr 18:141-152 (1995).

Larsen BR, Turrio-Baldassarri L, Nilsson T, Iacovella N, di Domenico A, Montagna M, Facchetti S. Toxic PCB congeners and organochlorine pesticides in Italian human milk. Ecotoxicol Environ Safety 28:1-13 (1994).

Larsen MC, Jefcoate CR. Phenobarbital induction of CYP2B1, CYP2B2, and CYP3A1 in rat liver: Genetic differences in a common regulatory mechanism. Arch Biochem Biophys 321:467-476 (1995).

Larsen MC, Brake PB, Parmar D, Jefcoate CR. The induction of five rat hepatic P450 cytochromes by phenobarbital and simmilarly acting compounds is regulated by a sexually dimorphic, dietary-dependent endocrine factor that is highly strain specific. Arch Biochem Biophys 315:24-34 (1994).

Larsson P. Transport of PCBs from aquatic to terrestrial environments by emerging chironomids. Environ Pollut 34A:283-289 (1984).

Lay JP, Klein W, Korte F. Beiträge zur ökologischen chemie C[1]. Ausscheidung, speicherung und metabolisierung von 2,4,6,2',4'-pentachlorbiphenyl-^{14}C nach langzeitfütterung an ratten. Chemosphere 3:161-168 (1975).

Lead WA, Steinnes E, Jones KC. Atmospheric deposition of PCBs to moss (Hylocomium splendens) in Norway between 1977 and 1990. Environ Sci Technol 30:524-530 (1996).

Letcher R, Norstrom R, Bergman Å. Geographical distribution and identification of methyl sulphone PCB and DDE metabolites in pooled polar bear (*Ursus maritimus*) adipose tissue from western hemisphere Arctic and subarctic regions. Sci Tot Environ 160:409-420 (1995).

Letcher RJ, Muir DCG, Norstrom RJ, Michaud R, Béland P. PCB biotransformation and methyl sulfone PCB metabolites in Beluga whale (*Delphinaptrerus leucas*) from the St. Lawrence River estuary and western Hudson Bay. Organohal Comp 33:290-295 (1997).

Lewis RG, Martin BE, Sgontz DL, Howes JE Jr. Measurement of fugitive atmospheric emissions of polychlorinated biphenyls from hazardous waste landfills. Environ Sci Technol 19:986-991 (1985).

Li M-H, Hansen LG. Enzyme induction and acute endocrine effects in prepubertal female rats receiving environmental PCB/PCDF/PCDD mixtures. Environ Hlth Persp 104:712-722 (1996a).

Li M-H, Hansen LG. Responses of prepubertal rats to environmental PCBs with high and low dioxin equivalencies. Fund Appl Toxicol 33:282-293 (1996b).

Li M-H, McKee MJ. Toxicokinetics of 2,2',4,4'- and 3,3',4,4'-tetrachlorobiphenyl congeners in the house cricket. Ecotoxicol Environ Safety 24:309-318 (1992).

Li M-H, Rhine C, Hansen LG. Hepatic enzyme induction and acute endocrine effects of 2,3,3',4',6-pentachlorobiphenyl in prepubertal female rats. Arch Environ Contam Toxicol 35:97-103 (1998).

Li M-H, Zhao Y-D, Hansen LG. Multiple dose toxicokinetic influence on the estrogenicity of 2,2',4,4',5,5'-hexachloro-biphenyl. Bull Environ Contam Toxicol 53:583-590 (1994).

Looser R, Ballschmiter K. Biomagnification of polychlorinated biphenyls (PCBs) in freshwater fish. Fresenius J Anal Chem 360:816-819 (1998).

Lorber M, Cleverly D, Schaum J, Phillips L, Schweer G, Leighton T. Development and validation of an air-to-beef food chain model for dioxin-like compounds. Science Tot Environ 156:39-65 (1994).

Lubet RA, Dragnev KH, Chauhan DP, Nims RW, Diwan BA, Ward JM, Jones CR, Rice JM, Miller MS. A pleiotropic response to phenobarbital-type enzyme inducers in the F344/NCr rat. Effects of chemicals of varied structure. Biochem Pharmacol 43:1067-1078 (1992).

Lucier GW, McDaniel OS, Schiller CM, Matthews HB. Structural requirements for the accumulation of chlorinated biphenyl metabolites in the fetal rat intestine. Drug Metab Dispos 6(5):584-590 (1978).

Lund BO, Bergman Å, Brandt I. Decreased pulmonary drug metabolism in mice treated with the PCB metabolite 4-methylsulphonyl-2,2',5,5'-tetrachlorobiphenyl. Toxicol Lett 32:261-267 (1986).

Lund BO, Bergman Å, Brunstrom B, Orberg J. Accumulation and long term effects of a PCB and DDE methyl sulfone mixture in the mink (*Mustela vison*). Organohal Compounds 33:360-365 (1997).

Luotamo M. Congener specific assessment of human exposure to polychlorinated biphenyls. Chemosphere 23:1685-1698 (1991).

Luotamo M, Jarvisalo J, Aitio A. Assessment of exposure to PCBs: Analysis of selected isomers in blood and adipose tissue. Environ Research 54:121-134 (1991).

Luotamo M, Patterson DG Jr, Needham LL, Aitio A. Concentrations of PCB congeners in sera from workers with past and present exposure. Chemosphere 27:171-177 (1993).

Lutz RJ, Dedrick RL. Physiological pharmacokinetic modeling of polychlorinated biphenyls. Environ Toxin Ser 1:111-131 (1987).

Lutz RJ, Dedrick RL, Matthews HB, Eling TE, Anderson MW. A preliminary pharmacokinetic model for several chlorinated biphenyls in the rat. Drug Metab Dispos 5:386-395 (1977).

Ma R, Cohen MB, Berenbaum MR, Schuler MA. Black swallowtail (*Papilio polyxenes*) alleles encode cytochrome P450s that selectively metabolize linear furanocoumarins. Arch Biochem Biophys 310:332-340 (1994).

Maack L, Sonzongni WC. Analysis of polychlorobiphenyl congeners in Wisconsin fish. Arch Environ Contam Toxicol 17:711-719 (1988).

MacDonald CR, Metcalfe CD, Balch GC, Metcalfe TL. Distribution of PCB congeners in seven lake systems: Interactions between sediment and food-web transport. Environ Toxicol Chem 12:1991-2003, 1993.

MacDonald CR, Norstrom RJ, Turle R. Application of pattern recognition techniques to assessment of biomagnification and sources of polychlorinated multicomponent pollutants, such as PCBs, PCDDs and PCDFs. Chemosphere 25:129-134 (1992).

MacDonald RW, Ikonomou MG, Paton DW. Historical inputs of PCDDs, PCDFs AND PCBs to a British Columbia interior lake: The effect of environmental controls on pulp mill emissions. Environ Sci Technol 32:331-337 (1998).

Machala M, Nezveda K, Irizar A, Bu-Abbas A, Ioannides C. Expression and inducibility of cytochrome P450 proteins in the liver of chick embryo. Arch Toxicol 71:57-63 (1996).

Mackay DJ, Shiu WY, Ma KC. *Illustrated handbook of physical-chemical properties and environmental fate for organic chemicals. Volume I: Monoaromatic hydrocarbons, chlorobenzenes and PCBs*; Lewis Publishers:Chelsea, MI, 1992; 324-697.

Maroni M, Colombi A, Cantoni S, Ferioli E, Foa V. Occupational exposure to polychlorinated biphenyls in electrical workers. I. Environmental and blood polychlorinated biphenyls concentrations. Brit J Indust Med 38:49-54 (1981).

Marsili L, Fossi MC, Casini S, Focardi S. PCB levels in bird blood and relationship to MFO responses. Chemosphere 33:699-710 (1996).

Matthews HB, Anderson MW. Effect of chlorination on the distribution and excretion of polychlorinated biphenyls. Drug Metab Dispos 3:371-380 (1975).

Matthews HB, Dedrick RL. Pharmacokinetics of PCBs. Ann Rev Pharmacol Toxicol 24:85-103 (1984).

Matthews HB, Tuey DB. The effect of chlorine position on the distribution and excretion of four hexachlorobiphenyl isomers. Toxicol Appl Pharmacol 53:377-388 (1980).

Maurel P. The CYP3 family. *In*: *Cytochromes P450: Metabolic and Toxicological Aspects* (Ioannides C, ed). Boca Raton: CRC Press (1996).

McCrady JK. Vapor phase 2,3,7,8-TCDD sorption to plant foliage--A species comparison. Chemosphere 28:207-216 (1994).

McFarland VA, Clarke JU. Environmental occurrence, abundance, and potential toxicity of polychlorinated biphenyl congeners: considerations for a congener-specific analysis. Environ Health Persp 81:225-239 (1989).

McKee MJ. Ecotoxicological evaluation of Area 9 Landfill at the Crab Orchard National Wildlife Refuge: Biological impact and residues. HWRIC RR-062, Waste Management and Research Center, Champaign, IL, 46 pp, 1992.

McKinney JD, Waller CL. Polychlorinated biphenyls as hormonally active structural analogues. Environ Health Perspec 102:290-297 (1994).

McLachlan MS. Digestive tract absorption of polychlorinated dibenzo-*p*-dioxins, dibenzofurans, and biphenyls in a nursing infant. Toxicol Appl Pharmacol 123:68-72 (1993).

McLean MR, Bauer U, Amaro AR, Robertson LW. Identification of catechol and hydroquinone metabolites of 4-monochlorobiphenyl. Chem Res Toxicol 9:158-164 (1996a).

McLean MR, Robertson LW, Gupta RC. Detection of PCB adducts by the ^{32}P-postlabeling technique. Chem Res Toxicol 9:165-171 (1996b).

Mes J, Davies DJ, Doucet J, Weber D, McMullen E. Specific polychlorinated biphenyl congener distribution in breast milk of Canadian women. Environ Technol 14:555-565 (1993).

Metcalf RL, Sanborn JR, Lu P-Y, Nye D. Laboratory model ecosystem studies of the degradation and fate of radiolabeled tri-, tetra-, and pentachlorobiphenyl compared with DDE. Arch Environ Contam Toxicol 3(2):151-165 (1975).

Miller DT, Condon SK, Kutzner S, Phillips DL, Krueger E, Timperi R, Burse VW, Cutler J, Gute DM. Human exposure to polychlorinated biphenyls in greater New Bedford, Massachusetts: a prevalence study. Arch Environ Contam Toxicol 20:410-416 (1991).

Mio T, Sumino K. Mechanism of biosynthesis of methyl sulfones from PCBs and related compounds. Environ Health Persp 59:129-135 (1985).

Mio T, Sumino K, Mizutani T. Sulfur-containing metabolites of 2,5,2',5'-tetrachlorobiphenyl, a major component of commercial PCB's. Chem Pharm Bull 24:1960-1985 (1976).

Mizutani T, Yamamoto K, Tajima K. Sulfur-containing metabolites of chlorobiphenyl isomers, a comparative study. J Agric Food Chem 26:862-866 (1978).

Moir D, Viau A, Chu I, Wehler EK, Morck A, Bergman Å. Tissue distribution, metabolism, and excretion of 2,4,4'-trichlorobiphenyl (CB-28) in the rat. Toxicol Ind Health 12(1):105-121 (1996).

Monosmith CL, Hermanson MH. Spatial and temporal trends of atmospheric organochlorine vapors in the central and upper Great Lakes. Environ Sci Technol 30:3464-3472 (1996).

Moore M, Mustain M, Daniel K, Chen I, Safe S, Zacharewski T, Gillesby B, Joyeux A, Balaguer P. Antiestrogenic activity of hydroxylated polychlorinated biphenyl congeners identified in human serum. Toxicol Appl Pharmacol 142:160-168 (1997).

Morales NM, Matthews HB. In vivo binding of 2,3,6,2',3',6'-hexachlorobiphenyl and 2,4,5,2',4',5'-hexachlorobiphenyl to mouse liver macromolecules. Chem Biol Interact 27:99-110 (1979).

Morales NM, Tuey DB, Colburn WA, Matthews HB. Pharmacokinetics of multiple oral doses of selected polychlorinated biphenyls in mice. Toxicol Appl Pharmacol 48:397-407 (1979).

Morrison HA, Gobas FAPC, Lazar R, Haffner GD. Development and verification of a bioaccumulation model for organic contaminants in benthic invertebrates. Environ Sci Technol 30:3377-3384 (1996).

Moza P, Scheuner I, Klein W, Korte F. Studies with 2,4',5-trichlorobiphenyl-^{14}C and 2,2',4,4'6-pentachlorobiphenyl-^{14}C in carrots, sugar beets and soil. J Agric Food Chem 27:1120-1124 (1979).

Moza P, Weisgerber I, Klein W, Korte F. Verteilung und metabolismus von 2,2'-dichlorbiphenyl-^{14}C in der höheren sumpfpflanze *Veronica beccabunga*. Chemosphere 5:217-222 (1973).

Mühlebach S, Bickel MH. Pharmacokinetics in rats of 2,4,5,2',4',5'-hexachlorobiphenyl, an unmetabolizable lipophilic model compound. Xenobiotica 11(4):249-257 (1981).

Muir DCG, Norstrom RI, Simon M. Organochlorine contaminants in Arctic marine food chains: Accumulation of specific polychlorinated biphenyls and chlordane-related compounds. Environ Sci Technol 22:1071-1079 (1988a).

Muir DCG, Omelchenko A, Grift NP, Savoie DA, Lockhard WL, Wilkinson P, Brunskill GJ. Spatial trends and historical deposition of polychlorinated biphenyls in Canadian midlatitude and Arctic lake sediments. Environ Sci Technol 30:3609-3617 (1996).

Muir DCG, Ford CA, Grift NP, Metner DA, Lockhard WL. Geographic variation of chlorinated hydrocarbons in burbot (*Lota lota*) from remote lakes and rivers in Canada. Arch Environ Contam Toxicol 19:530-542 (1990).

Muir DCG, Segstro MD, Hobson KA, Ford CA, Stewart REA, Olpinski S. Can seal eating explain elevated levels of PCBs and organochlorine pesticides in walrus blubber from Eastern Hudson Bay (Canada)? Environ Pollut 90:335-348 (1995).

Muir DCG, Segstro MD, Welbourn PM, Toom D, Eisenreich SJ, Macdonald CR, Whelpdale DM. Patterns of accumulation of airborne organochlorine contaminants in lichens from the upper Great Lakes region of Ontario. Environ Sci Technol 27:1201-1210 (1993).

Mullin CA, Scott JG (Eds). *Molecular Mechanisms of Insecticide Resistance: Diversity among Insects.* Amer Chem Soc, Washington, DC (1992).

Mullin MD, Pochini CM, McCrindle S, Romkes M, Safe SH, Safe LM. High-resolution PCB analysis: Synthesis and chromatographic properties of all 209 PCB congeners. Environ Sci Technol 18:468-476 (1984).

Murayama N, Shimada M, Yamazoe Y, Kato R. Difference in the suceptibility of two phenobarbital-inducible forms, P450IIB1 and P450IIB2, to thyroid hormone- and growth hormone-induced suppression in rat liver: Phenobarbital-inducible P450IIB2 suppression by thyroid hormone acting directly, but not through the pituitary system. Molec Pharmacol 39:811-817 (1991).

Murphy TJ, Formanski LJ, Brownawell B, Meyer JA. Polychlorinated biphenyl emissions to the atmosphere in the Great Lakes region. Municipal landfills and incinerators. Environ Sci Technol 19:942-946 (1985).

Nagayama J, Kiyohara C, Mohri N, Hirohata T, Haraguchi K, Masuda Y. Inhibitory effect of methylsulphonyl PCBs on aryl hydrocarbon hydroxylase activity. Chemosphere 18:701-708 (1989).

Nagayama J, Masuda Y. Effects of methylsulphonyl polychlorinated biphenyl homologues and 6,8-benzoflavone on aryl hydrocarbon hydroxylase activity. Fukuoka Acta Med 84:203-211 (1993).

Nagel SC, vom Saal FS, Thayer KA, Dhar MG, Boechler M, Welshons WV. Relative binding affinity-Serum modified access (RBA-SMA) assay predicts the relative in vivo bioactivity of the xenoestrogens Bisphenol A and octylphenol. Environ Health Perspect 105:70-76 (1997).

Namkung MJ, Yang HL, Hulla JE, Junchau MR. On the substrate specificity of cytochrome P450IIIA1. Mol Pharmacol 34:628-637 (1988).

Needham LL, Patterson DG Jr, Burse VW, Paschal DC, Turner WE, Hill RH Jr. Reference range data for assessing exposure to selected environmental toxicants. Toxicol Ind Health 12:507-513 (1996).

Nesaretnam K, Corcoran D, Dils RR, Darbre P. 3,4,3',4'-Tetrachlorobiphenyl acts as an estrogen *in vitro* and *in vivo*. Mol Endocrinol 10:923-936 (1996).

Ness DK, Schantz SL, Hansen LG. PCB congeners in the rat brain: Selective accumulation and lack of regionalization. J Toxicol Environ Hlth 43:453-468.

Ness DK, Schantz SL, Moshtaghian J, Hansen LG. Effects of perinatal exposure to specific PCB congeners on thyroid hormone concentrations and thyroid histology in the rat. Toxicol Lett 68:311-323 (1993).

Newsome WH, Andrews P. Organochlorine pesticides and polychlorinated biphenyl congeners in commercial fish from the Great Lakes. JAOAC Internat 76:707-710 (1993).

Newsome WH, Davies D. Determination of PCB metabolites in Canadian human milk. Chemosphere 33:559-565 (1996).

Newsome WH, Davies D, Doucet J. PCB and organochlorine pesticides in Canadian human milk - 1992. Chemosphere 30:2143-2153 (1995).

Nichols JW, Larsen CP, McDonald ME, Niemi GJ, Ankley GT. Bioenergetics-based model for accumulation of polychlorinated biphenyls by nestling tree swallows, *Tachycineta bicolor*. Environ Sci Technol 29:604-612 (1995).

Niimi AJ, Oliver BG. Biological half-lives of polychlorinated biphenyl (PCB) congeners in whole fish and muscle of rainbow trout (*Salmo gairdneri*). Can J Fish Aquat Sci 40:1388-1394 (1983).

Niimi AJ, Oliver BG. Distribution of polychlorinated biphenyl congeners and other halocarbons in whole fish and muscle among Lake Ontario salmonids. Environ Sci Technol 23:83-88 (1989).

Nims RW, Lubet RA. The CYP2B subfamily. *In. Cytochromes P450: Metabolic and Toxicological Aspects* (Ioannides C, ed). Boca Raton: CRC Press, 1996; 135-160.

Norback DH, Seymour JL, Knieriem KM, Peterson RE, Allen JR. Biliary metabolites of 2,5,2',5'-tetrachlorobiphenyl in the rat. Res Commun Chem Pathol Pharmacol 14:527-533 (1976).

Norén K, Lundén A, Pettersson E, Bergman Å. Methylsulfonyl metabolites of PCBs and DDE in human milk in Sweden, 1972-1992. Environ Health Persp 104:766-773 (1996).

Norén K, Meironyté D. Contaminants in Swedish human milk. Decreasing levels of organochlorine and increasing levels of organobromine compounds. Organohal Compounds 38:1-4 (1998).

Norstrom RJ, Muir DCG, Ford CA, Simon M, MacDonald CR, Beland P. Indications of P450 monooxygenase activities in Beluga (*Delphinapterus leucas*) and Narwhal (*Monodon monoceros*) from patterns of PCB, PCDD and PCDF accumulation. Mar Environ Res 34:267-272 (1992).

Norstrom RJ, Simon M, Muir DCG, Schweinsburg RE. Organochlorine contaminants in Arctic marine food chains: Identification, geographical distribution and temporal trends in polar bears. Environ Sci Technol 22:1063-1071 (1988).

Oakley GG, Devanaboyina U, Robertson LW, Gupta RC. Oxidative DNA damage induced by activation of polychlorinated biphenyls (PCBs): Implications for PCB-induced oxidative stress in breast cancer. Chem Res Toxicol 9:1285-1292 (1996).

Oberdorster E, Rittschof D, McClellan-Green P. Induction of cytochrome P450 3A by tributyltin in blue crabs, Callinectes sapidus. Toxicologist 36:136 (1997).

Øfjord GD, Puhakka JA, Feruson JF. Reductive dechlorination of Aroclor 1254 by marine sediment cultures. Environ Sci Technol 28:2287-2294 (1994).

Okey AB, Riddick DS, Harper PA. The Ah receptor: mediator of the toxicity of 2,3,7,8-tetrachlorodibenzo-p-dioxin (TCDD) and related compounds. Toxicol Lett 70:1-22 (1994).

Okey AB, Vella LM Elevated binding of 2,3,7,8-tetrachlorodibenzo-p-dioxin and 3-methylcholanthrene to the Ah receptor in hepatic cytosols from phenobarbital-treated rats and mice. Biochem Pharmacol 33:531-538 (1983).

Oliver BG, Charlton MN, Durham RW. Distribution, redistribution, and geochromology of polychlorinated biphenyl congeners and other chlorinated hydrocarbons in Lake Ontario sediments. Environ Sci Technol 23:200-208 (1989).

Ottar B. The transfer of airborne pollutants to the Arctic region. Atmos Environ 15:1439-1445 (1981)

Owens IS. Genetic regulation of UDP-glucuronosyl-transferase induction by polycyclic aromatic compounds in mice. J Biol Chem 252:2827-2833 (1985).

Paine JM, McKee MJ, Ryan ME. Toxicity and bioaccumulation of soil PCBs in crickets: Comparison of laboratory and field studies. Environ Toxicol Chem 12:2097-2103 (1993).

Pang S, Cao JQ, Katz BH, Hayes CL, Sutter TR, Spink DC. Inductive and inhibitory effects of non-*ortho*-substituted polychlorinated biphenyls on estrogen metabolism and human cytochromes P450 1A1 and 1B1. Biochem Pharmacol, in press (1999).

Panshin SY, Hites RA. Atmospheric concentrations of polychlorinated biphenyls at Bloomington, Indiana. Environ Sci Technol 28:2008-2013 (1994).

Parham FM, Kohn MC, Matthews HB, DeRosa C, Portier CJ. Using structural information to create physiologically based pharmacokinetic models for all polychlorinated biphenyls. I. Tissue:Blood Partition Coefficients. Toxicol Appl Pharmacol 144:340-347 (1997).

Parham FM, Portier CJ. Using structural information to create physiologically based pharmacokinetic models for all polychlorinated biphenyls. II. Rates of Metabolism. Toxicol Appl Pharmacol 151:110-116 (1998).

Patandin S, Dganelie PC, Weisglas-Kuperus N, Sauer PJJ. Exposure to PCBs, PCDDs and PCDFs through breast milk compared with long-term dietary exposure. Organohal Compounds 38:214A-214F (1998).

Paterson S, Mackay D, Bacci E, Calamari D. Correlation of the equilibrium and kinetics of leaf-air exchange of hydrophobic organic chemicals. Environ Sci Technol 25:866-871 (1991).

Peakall DB, Lincer JL, Bloom SE. Embryonic mortality and chromosomal alterations caused by Aroclor 1254 in ring doves. Environ Health Perspect 1:103-104 (1972).

Pearson RF, Hornbuckle KC, Eisenreich SJ, Swackhamer DL. PCBs in Lake Michigan water revisited. Environ Sci Technol 30:1429-1436 (1996).

Pessah IN, Hansen LG, Ta TA, Mai E, Wong PW. Noncoplanar PCBs can discriminate between genetic isoforms of microsomal immunophilin/CA2+ channel complexes. Toxicologist 48:277 (1999).

Phillips LJ, Birchard G. Regional variations in human toxic exposure in the USA: an analysis based on the National Adipose Tissue Survey. Arch Environ Contam Toxicol 21:159-168 (1991).

Platonow NS, Geissinger HD. Distribution and persistence of PCBs (Aroclor 1254) in growing piglets. Vet Record 93:287 (1973).

Poland A, Knutson JC. 2,3,7,8-tetrachlorodibenzo-p-dioxin and related halogenated aromatic hydrocarbons: examination of the mechanisms of toxicity. Ann Rev Pharmacol Toxicol 22:517-554 (1982).

Polzin KL, Speer KG, Toole JM, Schmitt RW. Intense mixing of antarctic bottom water in the equatorial Atlantic Ocean. Nature 380:54-56 (1996).

Porte C, Albaigés J. Bioaccumulation patterns of hydrocarbons and polychlorinated biphenyls in bivalves, crustaceans, and fishes. Arch Environ Contam Toxicol 26:273-281 (1993).

Portier RJ, Fujisaki K. Enhanced biotransformation and biodegradation of PCBs in the presence of aminopolysaccharides. *In* Adams WJ, Chapman GA, Landis WG (eds) *Aquatic Toxicology and Hazard Assessment: 10th Volume.* ATSTM STP 971, Philadelphia, pp. 517-527 (1988).

Poster DL, Baker JE. Influence of submicron particles on hydrophobic organic contaminants in precipitation. 1. Concentrations and distributions of polycyclic aromatic hydrocarbons and polychlorinated biphenyls in rainwater. Environ Sci Technol 30:341-348 (1996).

Preston BD, Miller JA, Miller EC. Non-arene oxide aromatic ring hydroxylation of 2,2',5,5'-tetrachlorobiphenyl as the major metabolic pathway catalyzed by phenobarbital-induced rat liver microsomes. J Biol Chem 258:8304-8311 (1984).

Pruell RJ, Rubinstein NI, Taplin BK, LiVolsi JA, Bowen RD. Accumulation of polychlorinated organic contaminants from sediment by three benthic marine species. Arch Environ Contam Toxicol 24:290-297 (1993).

Puttmann M, Mannschreck A, Oesch F, Robertson LW. Chiral effects in the induction of drug-metabolizing enzymes using synthetic atropisomers of polychlorinated biphenyls (PCBs). Biochem Pharmacol 38:1345-1352 (1989).

Quensen JF III, Tiedje JM, Boyd SA. Reductive dechlorination of PCBs by anaerobic microorganisms from sediments. Science 242:752-754 (1988).

Rappe C. Chemical behavior of pesticides pp. 29-37 *In*: *Ecological Problems of the Circumpolar Area*. Norbottens Museaum, Lulea, Sweden (1974).

Reeder AL, Foley GL, Nichols DK, Hansen LG, Faeh S, Eisold J, Wheeler MB, Warner R, Murphy JE, Beasley VR. Forms and prevalence of intersexuality,and effects of environmental contaminants on sexuality in cricket frogs (*Acris crepitans*). Environ Health Persp 106:261-266 (1998).

Reich S, Jimenez B, Marsili L, Herandez LM, Schurig V, Gonzalez MJ. Enantiomeric ratios of chiral PCBs in stripped dolphins (*Stenella coeruleoalba*) from the Mediterranean Sea. Organohal Compounds 35:335-338 (1998).

Rhee KS, Plapp FW Jr. Polychlorinated biphenyls (PCBs) as inducers of microsomal enzyme activity in the house fly. Arch Environ Contam Toxicol 1:182-192 (1972).

Robens J, Anthony HD. Polychlorinated biphenyl contamination of feeder cattle. J Amer Veter Med Assoc 177:613-615 (1980).

Rodman LE, Shedlofsky SI, Mannschreck A, Puttmann M, Swim AT, Robertson LW. Differential potency of atropisomers of PCBs on cytochrome P450 induction and uroporphyrin accumulation in the chick embryo hepatocyte culture. Biochem Pharmacol 41:915-922 (1991).

Rothman N, Cantor KP, Blair A, Bush D, Brock JW, Helzlsouer K, Zahm SH, Needham LL, Pearson GR, Hoover RN, Comstock GW, Strickland PT. A nested case-control study of non-Hodgkin lymphoma and serum organochlorine residues. Lancet 350:240-244 (1997).

Sabljic A, Gusten H, Schonherr J, Riederer M. Modelling plant uptake of airborne organic chemicals. 1. Plant cuticle/water partitioning and molecular connectivity. Environ Sci Technol 24:1321-1326 (1990)

Safe SH. Polychlorinated biphenyls (PCBs): Environmental impact, biochemical and toxic responses, and implications for risk assessment. Crit Rev Toxicol 24(2):87-149 (1994).

Safe S. Polyhalogenated aromatics: uptake, disposition and metabolism, in *Halogenated Biphenyls, Naphthalenes, Dibenzodioxins and Related Compounds*, Kimbrough RD, Jensen AA, eds. Elsevier-North Holland, Amsterdam (1989).

Safe S, Hutzinger O, Jones D. The mechanism of chlorobiphenyl metabolism. J Agric Food Chem 23:851-853 (1975).

Safe S, Bandiera S, Sawyer T, Robertson L, Safe L, Parkinson A, Thomas PE, Ryan DE, Reik LM, Levin W, Denomme MA, Fujita T. PCBs. Structure-function relationships and mechanism of action. Environ Health Persp 60:47-56 (1985).

Safe S, Safe L, Mullin M. Polychlorinated biphenyls: Congener-specific analysis of a commercial mixture and a human milk extract. J Agric Food Chem 33:24-29 (1985).

Saghir SA, Hansen LG. Distribution and metabolism of 2,5,2'-trichlorobiphenyl in houseflies (*Musca domestica* L). Bull Environ Contam Toxicol 49:839-846 (1992).

Saghir SA, Hansen LG. Toxicities of 2,2',4,4'- and 3,3',4,4'-tetrachlorobiphenyls to house flies at different ages and enzyme levels. Arch Environ Contam Toxicol 28:436-442 (1995).

Saghir SA, Hansen LG. Toxicity and tissue distribution of 2,2',4,4'- and 3,3',4,4'-tetrachlorobiphenyls in houseflies. Ecotoxicol Environ Saf 42:in press (1999).

Saghir SA, Koritz GD, Hansen LG. Toxicokinetics of 2,2',4,4'- and 3,3',4,4'-tetrachlorobiphenyl in house flies following topical administration. Pestic Biochem Physiol 49:94-113 (1994).

Saghir SA, Koritz GD, Hansen LG. Toxicokinetics of 2,2',5-trichlorobiphenyl in house flies following topical administration. Pestic Biochem Physiol 46:107-119 (1993).

Saghir SA, Koritz GD, Hansen LG. Short-term distribution, metabolism, and excretion of 2,2',5-tri-, 2,2',4,4'-tetra- and 3,3'4,4'-tetrachlorobiphenyls in prepubertal rats. Arch Environ Contam Toxicol 36:213-220 (1999).

Saghir SA, Schaeffer DJ, Hansen LG. Uptake and metabolism of [14-C]-2,2',5-trichlorobiphenyl in *Dugesia dorotocephala*. Fresenius Environ Bull 4:31-34 (1995).

Sandau CD, Ayotte P, Dewailly E, Bergman Å, Klasson-Wehler E, Norstrom RJ. Analysis of hydroxylated metabolites of PCBs (OH-PCBs) in whole blood from Canadian Inuit. Organohal Compounds 38:29-32 (1998).

Sanders G, Hamilton-Taylor J, Jones KC. PCB and PAH dynamics in a small rural lake. Environ Sci Technol 30:2958-2966 (1996).

Schantz MM, Paris RM, Kurz J, Ballschmiter K, Wise SA. Comparison of methods for the gas-chromatographic determination of PCB congeners and chlorinated pesticides in marine reference materials. Fresenius J Anal Chem 346:766-778 (1993).

Schantz SL, Seo BW, Wong PW, Pessah IN. Long-term effects of developmental exposure to 2,2',3,5',6-pentachlorobiphenyl (PCB 95) on locomotor activity, spatial learning and memory and brain ryanodine receptors. Neurotoxicol 18:457-468 (1997).

Schmoldt A, Herzberg W, Benthe HF. On the inhibition of microsomal drug metabolism by polychlorinated biphenyls (PCB) and related phenolic compounds. Chem Biol Interact 16:191-200 (1977).

Schnellmann RG, Putnam CW, Sipes IG. Metabolism of 2,2',3,3',6,6'-hexachlorobiphenyl and 2,2',4,4',5,5'-hexachlorobiphenyl by human hepatic microsomes. Biochem Pharmacol 32:3233-3239 (1983).

Schnellmann RG, Volp RF, Putnam CW, Sipes IG. The hydroxylation, dechlorination, and glucuronidation of 4,4'-dichlorobiphenyl by human hepatic microsomes. Biochem Pharmacol 33:3503-3509 (1984).

Schreitmüller J, Ballschmiter K. Levels of polychlorinated biphenyls in the lower troposphere of the North- and South-Atlantic Ocean. Fresenius J Anal Chem 348:226-239 (1994).

Schulte E, Malsich R. Berechnung der wahren PCB-Gehalt in Umweltproben. I. Ermittlung der Zussammensetzung Zweier technischer PCB-Gemische. Fres Z Anal Chem 1983;314:545-51.

Schultz TW. Letter to The Editor (re: Genoni, 1997) Ecotoxicol Environ Safety 38:336-338 (1997).

Schwartz TR, Stalling DL, Rice CL. Are polychlorinated biphenyl residues adequately described by Aroclor mixture equivalents? Isomer-specific principal components analysis of such residues in fish and turtles. Environ Sci Technol 21:72-76 (1987).

Seegal RF, Brosch KO. The co-planar PCB congener 3,3',4,4',5-pentachlorobiphenyl is uterotropic in prepubertal rats. Toxicologist 48:276 (1999).

Serabjit-Singh CJ, Albro PW, Robertson IGC, Philpot RM. Interactions between xenobiotics that increase or decrease the levels of cytochrome P-450 isozymes in rabbit lung and liver. J Biol Chem 258:12827-12834 (1983).

Shain W, Bush B, Seegal R. Neurotoxicity of polychlorinated biphenyls: Structure-Activity relationships of individual congeners. Toxicol Appl Pharmacol 111:33-42 (1991).

Shargel L, Yu ABC. *Applied Biopharmaceutics and Pharmacokinetics*, Norwalk, Appleton and Lange, 625 pp. (1993).

Sheiner LB. The population approach to pharmacokinetic data analysis: rationale and standard data analysis methods. Drug Metab Rev 15:153-171 (1984).

Shimada T, Sato R. Covalent binding *in vitro* of polychlorinated biphenyls to microsomal macromolecules. Biochem Pharmacol 27:585-? (1978).

Shimada T, Sawabe Y. Comparative studies on distribution and covalent tissue binding of 2,4,2',4'-tetrachlorobiphenyl isomers in the rat. Arch Toxicol 55:182-185 (1984).

Shimada T, Sawabe Y, Nakano Y. Interaction of 3,4,3',4'-tetrachlorobiphenyl metabolites formed by cytochrome P-450 in vitro with rat erythrocytes. Arch Toxicol 58:20-26 (1985).

Shimada T. Metabolic activation of [^{14}C]polychlorinated biphenyl mixtures by rat liver microsomes. Bull Environ Contam Toxicol 16:25-31 (1976).

Shimada T. Lack of correlation between formation of reactive metabolites and thymic atrophy caused by 3,4,3',4'-tetrachlorobiphenyl in C57BL/6N mice. Arch Toxicol 59:301-306 (1987).

Simcik MF, Franz TP, Zhang H, Eisenreich SJ. Gas-particle partitioning of PCBs and PAHs in the Chicago urban and adjacent coastal atmosphere: States of equilibrium. Environ Sci Technol 32:251-257 (1998).

Sipes IG, Schnellmann RG. Biotransformation of PCBs: Metabolic pathways and mechanisms. Environ Toxin Ser 1:97-110 (1987).

Sipes IG, Slocum ML, Chen G, Carter DE. 2,3,6,2',3',6'-Hexachlorobiphenyl: Distribution, metabolism, and excretion in the dog and the monkey. Toxicol Appl Pharmacol 65:264-272 (1982a).

Sipes IG, Slocum ML, Perry DF, Carter DE. 2,4,5,2',4',5'-Hexachlorobiphenyl: Distribution, metabolism, and excretion in the dog and the monkey. Toxicol Appl Pharmacol 65:264-272 (1982b).

Sissons D, Welti D. Structural identification of polychlorinated biphenyls in commercial mixtures by gas liquid chromatography, nuclear magnetic resonance and mass spectrometry. J Chromatogr 60:15-32 (1971).

Smith D. Are PCBs in the Great Lakes approaching a "new equilibrium"? Environ Sci Technol 29:42A-46A (1995).

Sonzogni W, Maack L, Gibson T, Degenhardt D, Anderson H, Fiore B. Polychlorinated biphenyl congeners in blood of Wisconsin sport fish consumers. Arch Environ Contam Toxicol 20:56-60 (1991).

Soontornchat S, Li M-H, Cooke PS, Hansen LG. Toxicokinetic and toxicodynamic influences on endocrine disruption by polychlorinated biphenyls. Environ Health Perspec 102:568-571 (1994).

Spink DC, Hayes CL, Young NR, Christou M, Sutter TR, Jefcoate CR, Gierthy JF. The effects of 2,3,7,8-tetrachlorodibenzo-*p*-dioxin on estrogen metabolism in MCF-7 breast cancer cells: Evidence for induction of a novel 17β-estradiol 4-hydroxylase. J Steroid Biochem Molec Biol 51:251-258 (1994).

Spink BC, Fasco MJ, Gierthy JF, Spink DC. 12-*O*-Tetradecanoylphorbol-13-acetate upregulates the Ah receptor and differentially alters CYP1B1 and CYP1A1 expression in MCF-7 breast cancer cells. J Cellular Biochem, in press (1998).

Stalling DL, Huckins JN, Petty JD, Johnson JL, Sanders HO. An expanded approach to the study and measurement of PCBs and selected planar halogenated aromatic environmental pollutants. Ann NY Acad Sci 320:48-59 (1979).

Stegeman JJ. Benzo[a]pyrene oxidation and microsomal enzyme activity in the mussel (*Mytilus edulis*) and other bivalve mollusc species from the Western North Atlantic. Marine Biology 89:21-30 (1985).

Stegeman JJ, Lech JJ. Cytochrome P-450 monooxygenase systems in aquatic species: Carcinogen metabolism and biomarkers for carcinogen and pollutant exposure. Environ Health Perspect 90:101-109 (1991).

Stegeman JJ, Kloepper-Sams PJ, Farrington JW. Monooxygenase induction and chlorobiphenyls in the deep-sea fish Coryphaenoides armatus. Science 231:1287-1289 (1986).

Stegeman JJ, Woodin BR, Singh H, Oleksiak MF, Celander M. Cytochromes P450 (CYP) in tropical fishes: Catalytic activities, expression of multiple CYP proteins and high levels of microsomal P450 in liver of fishes from Bermuda. Comp Biochem Physiol 116C:61-75 (1997).

Steimer J-L, Vozeh S, Racine-Poon A, Holford N, O'Neill R. The population approach: Rationale, methods, and applications in clinical pharmacology and drug development. *In* Wellington PG, Balant LP (eds) *Pharmacokinetics of Drugs*. New York, Springer-Verlag (1994) pp 405-451.

Storr-Hansen E. Coplanar polychlorinated biphenyl congeners: Development of an analytical method and its application to samples from Danish harbour seal (*Phoca vitulina*). PhD Thesis. National Environmental Research Institute, Roskilde, Denmark, 176 pp (1992).

Storr-Hansen E, Koritz GD, Hansen LG. Comparative cuticle penetration and whole-body elimination of PCB congeners in the house fly. Nordic Symposium on Organic Environmental Chemicals, Joensus, Finland, pp. 77-79 (1989).

Storr-Hansen E, Spliid H. Distribution patterns of polychlorinated biphenyl congeners in harbour seal (*Phoca vitulina*) tissues: Statistical analysis. Arch Environ Contam Toxicol 25:328-345 (1993).

Sundlof SF, Hansen LG, Koritz GD, Sundlof SM. The pharmacokinetics of hexachlorobenzene in male beagles. Distribution, excretion, and pharmacokinetic model. Drug Metab Dispos 10(4):371-381.

Sundström G, Jansson B. The metabolism of 2,2',3,5',6-pentachlorobiphenyl in rats, mice and quails. Chemosphere 6:361-370 (1975).

Sundström G, Hutzinger O, Safe S. The metabolism of chlorobiphenyls--A review. Chemosphere 5:267-298 (1976).

Swackhamer DL, Hites RA. Occurrence and bioaccumulation of organochlorine compounds in fishes from Siskiwit Lake, Isle Royale, Lake Superior. Environ Sci Technol 22:543-548 (1988).

Swackhamer DL, Skoglund RS. Bioaccumulation of PCBs by algae: Kinetics versus equilibrium. Environ Toxicol Chem 12:831-838 (1993).

Swackhammer DL, Armstrong DE. Estimation of the atmospheric and nonatmospheric contributions and losses of polychlorinated biphenyls for Lake Michigan on the basis of sediment records of remote lakes. Environ Sci Technol 20:879-883 (1986).

Tanabe S, Hidaka H, Tatsukawa R. PCBs and chlorinated hydrocarbon pesticides in Antarctic atmosphere and hydrosphere. Chemosphere 12:277-288 (1983).

Tanabe S, Kannan K, Tabucanon MS, Siriwong C, Ambe Y, Tatsukawa R. Organochlorine pesticide and polychlorinated biphenyl residues in foodstuffs from Bangkok, Thailand. Environ Pollut 72:191-203 (1991).

Tanabe S, Kannan N, Wakimoto T, Tatsukawa R, Okamoto T, Masuda Y. Isomer-specific determination and toxic evaluation of potentially hazardous coplanar PCBs, dibenzofurans and dioxins in the tissues of "yusho" PCB poisoning victim and in the causal oil. Toxicol Environ Chem 24:215-231 (1990).

Tatsukawa R, Tanabe S. Geochemical and biochemical behavior of PCBs in the open ocean environment. *In* Barros MC, Konemann H, Visser R (eds) *PCB Seminar*, The Hague, The Netherlands, Ministry of Housing, Physical Planning and Environment, pp. 99-118 (1984).

Tehseen WM, Hansen LG, Schaeffer DJ. Polychlorinated biphenyl (PCB) congener effects on the longevity of the housefly. Bull Environ Contam Toxicol 48:101-107 (1992).

Thomas G, Sweetman A, Lead W, Jones KC. Air-pasture transfers of PCBs. Organohal Compounds 33:259-263 (1997).

Tiedje JM, Quensen JF III, Chee-Sanford J, Schimel JP, Boyd SA. Microbial reductive dechlorination of PCBs. Biodegradation 4:231-240 (1993).

Tillitt DE, Gale RW, Meadows JC, Zajicek JL, Peterman PH, Heaton SN, Jones PD, Bursian SJ, Kubiak TJ, Giesy JP, Aulerich RJ. Dietary Exposure of mink to carp from Saginaw Bay. 3. Characterization of dietary exposure to planar halogenated hydrocarbons, dioxin equivalents and biomagnification. Environ Sci Technol 30:283-291 (1996).

Treon JF, Cleveland FP, Cappel JW, Atchley RW. The toxicity of the vapours of Aroclor 1242 and Aroclor 1254. Am Ind Hyg Assoc J 17:204-213 (1956).

Troisi GM, Mason CF. Cytochromes P450, P420 and mixed-function oxidases as biomarkers of polychlorinated biphenyl (PCB) exposure in harbour seals (*Phoca vitulina*). Chemosphere 35:1933-1946 (1997).

Tuinstra LGMTh, Huisman M, Boersma ER. The Dutch PCB/dioxin study contents of dioxins, planar and other PCBs in human milk from the Rotterdam and Groningen area. Chemosphere 29:2267-2277 (1994).

Tuinstra LGMTh, Roos AH, Griepink B, Wells DE. Interlaboratory studies on the determination of selected chlorobiphenyl congeners with capillary gas chromatography using splitless- and on-column injection techniques. J High Resolut Chromatogr Commun 8:475-480 (1985).

Tuinstra LGMTh, Traag WA. Capillary gas chromatographic-mass spectrometric determination of individual chlorobiphenyls in technical Aroclors. J Assoc Offic Anal Chem 66:708-717 (1983).

Turner JC. Chlorinated hydrocarbon residues in perianal fat of desert bighorn sheep. Bull Environ Contam Toxicol 7:23-31 (1978).

Turrio-Baldassarri L, di Domenico A, Fulgenzi A, La Rocca C, Iacovella N, Rodriguez F, Volpi, F. Influence of relative response factor in the determination of organic microcontaminants with isotopically labeled standards. Mikrochim Acta 123:45-53 (1996).

Turrio-Baldassarri L, di Domenico A, Iacovella N, La Rocca C, Mediati MG, Rodriguaz F. PCB congener profile and contamination levels of Italian National and regional diets. Organohal Compounds 26:121-124 (1995).

Turrio-Baldassarri L, D'Augostine O, De Felip E, di Domenico A, Fulgenzi AR, Iacovella N, La Rocca C, Rodriguez F, Volpi F. The ratio of 2,2',3,4',5',6-hexachlorobiphenyl (149) to 2,2',4,4',5,5'-hexachlorobiphenyl (153) as a possible degradational and analytical indicator. Microchem J 55:1-11 (1997).

Turrio-Baldassarri L, di Domenico A, Fulgenzi AR, Iacovella N, La Rocca C. Levels of polychlorobiphenyl congeners in mean diet samples from different Italian areas. Organohal Compounds 38:195-198 (1998).

USEPA. Guidance for Assessing Chemical Contaminant Data for use in Fish Advisories. Vol 1. Fish Sampling and Analysis. Environmental Protection Agency, Office of Water (4305), EPA 823-R-93-002 (1993).

USEPA. Methodology of assessing health risks associated with multiple exposure pathways to combustion emissions. Chapter 9. Breast Milk Pathway. Draft 04/01/97; Revised 01/26/98. National Center of Environmental Assessment, Cincinnati, OH (1998).

van der Oost, R, van Gastel L, Worst D, Hanraads, M, Satumalay K, van Schooten F-J, Heida H, Vermeulen NPE. Biochemical markers in feral roach (*Rutilus rutilus*) in relation to the bioaccumulation of organic trace pollutants. Chemosphere 29:801-817 (1994).

Van Birgelen APJM, Van der Kolk J, Poiger H, Van den Berg M, Brouwer A. Interactive effects of 2,2',4,4',5,5'-hexachlorobiphenyl and 2,3,7,8-tetrachlorodibenzo-p-dioxin on thyroid hormone, vitamin A, and vitamin K metabolism in the rat. Chemosphere 25:1239-1244 (1992).

van Vliet T. Polychlorinated Biphenyls (PCBs): Interactions between different congeners in *in vitro* enzyme induction assays. IMM-rapport 3/90, Karolinska Institutet, Stockholm, 48 pp. (1990).

Vartiainen T, Saarikoski S, Jaakkola JJ, Tuomisto J. PCDD, PCDF, and PCB concentrations in human milk from two areas in Finland. Chemosphere 34:2571-2583 (1997).

Vermette S, Willet M, Cochran J. Air concentrations of PCBs and metals at Crab Orchard National Wildlife Center. HWRIC RR-063. Waste Management and Research Center, Champaign, IL 57 pp (1995).

Vetter W, Luckas B. Calculation algorithm for the retention times of polychlorinated biphenyls and the control of its efficiency on a non-polar capillary column. J Chromatogr A 699:173-182 (1995).

Vetter W, Natzeck C, Luckas B, Heidemann G, Kiabi B, Karami M. Chlorinated hydrocarbons in the blubber of a seal (*Phoca caspica*) from the Caspian Sea. Chemosphere 30:1685-1696 (1995).

Viganò L, Galassi S, Arillo A. Bioconcentration of polychlorinated biphenyls (PCBs) in Rainbow trout caged in the River Po. Ecotoxicol Environ Safety 28:287-297 (1994).

Villeneuve DC. The effect of food restriction on the redistribution of hexachlorobenzene in the rat. Toxicol Appl Pharmacol 31:313-319 (1975).

Viluksela M, Stahl BU, Birnbaum LS, Rozman KK. Subchronic/chronic toxicity of 1,2,3,4,6,7,8-heptachlorodibenzo-p-dioxin (HpCDD) in rats. II. Biochemical effects. Toxicol Appl Pharmacol 146:217-226 (1997).

vom Saal FS, Timms BG, Montano MM, Palaza P, Thayer KA, Nagel SC, Dhar MD, Ganjam VK, Parmigiani S, Welshons WV. Prostate enlargement in mice due to fetal exposure to low doses of estradiol or diethylstilbestrol and opposite effects at high doses. Proc Natl Acad Sci USA 94:2056-2061 (1997).

vom Saal FS, Welshons WV, Hansen LG. Organochlorine residues and breast cancer. Letter to the Editor. N Engl J Med 338:988 (1998).

Voorman R, Aust SD. TCDD (2,3,7,8-tetrachlorodibenzo-p-dioxin) is a tight binding inhibitor of cytochrome P450d. J Biochem Toxicol 4:105-109 (1989).

Wallace JC, Basu I, Hites RA. Sampling and analysis artifacts caused by elevated indoor air polychlorinated biphenyls concentrations. Environ Sci Technol 30:2730-2734 (1996).

Wallace JC, Hites RA. Diurnal variations in atmospheric concentrations of polychlorinated biphenyls and endosulfan: Implications for sampling protocols. Environ Sci Technol 30:444-446 (1996).

Waller CL, Minor DL, McKinney JD. Using three-dimensional quantitative structure-activity relationships to examine estrogen receptor binding affinities of polychlorinated hydroxybiphenyls. Environ Health Perspec 103:702-707 (1995).

Wania F, Mackay D. Global fractionation and cold condensation of low volatility organochlorine compounds in polar regions. Ambio 22:10-18 (1993).

Wania F, Mackay D. Tracking the distribution of persistent organic pollutants. Environ Sci Technol 390A-396A (1996).

Waxman DJ, Azaroff L. Phenobarbital induction of cytochrome P-450 gene expression. Biochem J 281:577-592 (1992).

Webb RG, McCall AC. Identities of polychlorinated biphenyl isomers in Aroclors. J Assoc Offic Anal Chem 55:746-752 (1972).

Webb RG, McCall AC. Quantitative PCB standards for electron capture gas chromatography. J Chromatogr Sci 11:366-373 (1973).

Weistrand C, Norén K. Methylsulfonyl metabolites of PCBs and DDE in human tissues. Environ Health Persp 105:644-649 (1997).

Wellington Laboratories. Analytical Report to BOVAR Environmental, December 16, 1996. Calgary, Alberta.

Wells DE, Echarri I. Determination of chlorobiphenyls, with the separation of non-ortho, mono-ortho and di-ortho congeners in fish and sea mammals. Analyt. Chim. Acta 286:431-449 (1994).

White INH, Davies A, Smith LL, Dawson S, de Matteis F. Induction of CYP2B1 and 3A1, and associated monooxygenase activities by tamoxifen and certain analogues in the livers of female rats and mice. Biochem Pharmacol 45:21-30 (1993).

White RD, Shea D, Solow AR, Stegeman JJ. Induction and post-transcriptional suppression of hepatic cytochrome P450 1A1 by 3,3',4,4'-tetrachlorobiphenyl. Biochem Pharmacol 53:1029-1040 (1997).

Whittem T. Pharmacokinetics and milk discard times of pillomycin after intramammary infusion: A population approach. J Veterin Pharmacol Therap, in press (1999).

Whittem EL, Hansen LG, Schaeffer DJ. Neuroendocrine effects in children from breast milk pathway exposure to PCB mixtures - hypothesis creation. 2. Biomathematical models. Report to USEPA -NCEA (Cincinnati), Sept 27, 1998.

Wilken A, Bock C, Bokern M, Harms H. Metabolism of different PCB congeners in plant cell cultures. Environ Toxicol Chem 14:2017-2022 (1995).

Wilson DW, Hansen LG. Changes in kinetic characteristics of sheep liver microsomes following induction and dietary stress. Toxicol Appl Pharmacol 31:114-119 (1975).

Winter S, Streit B. Organochlorine compounds in a three-step terrestrial food chain. Chemosphere 24(12):1765-1774 (1992).

Wolff MS, Fischbein A, Selikoff IJ. Changes in PCB serum concentrations among capacitor manufacturing workers. Environ Research 59:202-216 (1992).

Wolff MS, Thornton J, Fischbein A, Lilis R, Selikoff IJ. Disposition of polychlorinated biphenyl congeners in occupationally exposed persons. Toxicol Appl Pharmacol 62:294-306 (1982).

Wolff MS, Toniolo PG, Lee EW, Rivera M, Dubin N. Blood levels of organochlorine residues and risk of breast cancer. J Natl Canc Inst 85:648-652 (1993).

Wolff MS, Camann D, Gammon M, Stellman SD. Proposed PCB congener groupings for epidemiological studies. Environ Health Perspect 105:13-14 (1997).

Wong PW, Pessah IN. Ortho-substituted PCBs alter calcium regulation by a ryanodine receptor-mediated mechanism: Structural specificity toward skeletal- and cardiac-type microsomal calcium release channels. Molec Pharmacol 49:740-751 (1996).

Wong PW, Pessah IN. Non-coplanar PCB 95 alters microsomal Ca^{+2} transport by an immunophilin FKBP12-dependent mechanism. Molec Pharmacol 51:693-702 (1997).

Wyndham C, Devenish J, Safe S. The in vitro metabolism, macromolecular binding and bacterial mutagenicity of 4-chlorobiphenyl, a model PCB substrate. Res Commun Chem Pathol Pharmacol 15:563-570 (1976).

Yakushiji T, Watanabe I, Kuwabara K, Yoshida S, Koyama K, Kunita N. Levels of polychlorinated biphenyls (PCB) and organochlorine pesticides in human milk and blood collected in Osaka Prefecture from 1972-1977. Int Arch Occup Environ Health 43:1-15 (1979).

Yamamoto H, Yoshimura H. Metabolic studies on Polychlorinated biphenyls. III. Complete structure and acute toxicity of the metabolite of 2,4,3',4'-tetrachlorobiphenyl. Chem Pharm Bull (Japan) 21:2237-2238 (1973).

Yamamoto H, Yoshimura H. A novel route of excretion of 2,4,3',4'-tetrachlorobiphenyl in rats. Bull Environ Contam Toxicol 13(6):681-688 (1975).

Yoshimura H, Yamamoto H, Saeki S. Metabolic studies on Polychlorinated biphenyls. II. Metabolic fate of 2,4,3',4'-tetrachlorobiphenyl in rats. Chem Pharm Bull (Japan) 21:231-236 (1973).

Zell M, Ballschmiter K. Baseline studies of the global pollution. III. Trace analysis of polychlorinated biphenyls (PCB) by ECD glass capillary gas chromatography in environmental samples of different trophic levels. Fresenius Z Anal Chem 304:337-349 (1980).

Zell M, Neu HJ, Ballschmiter K. Single component analysis of polychlorinated biphenyl (PCB)- and chlorinated pesticide residues in marine fish samples. Fresenius Z Anal Chem 292:97-107 (1978).

Ziprin RL, Elissalde MH, Clark DE, Wilson RD. Absorption of polychlorinated biphenyl by the ovine lymphatic system. Veterin Hum Toxicol 22:305-398 (1980).

Zuccato E, Calvarese S, Mariani G, Mangiapan S, Grasso P, Guzzi A, Fanelli R. Level, sources and toxicity of polychlorinated biphenyls in the Italian diet. Chemosphere, in press (1999).

Appendix Table 1. Nomenclature of Polychlorinated Biphenyls

IUPAC No.	Compound	CAS No.	Melting Pt.
	Biphenyl	92-52-4	71
	Monochlorobiphenyl	27323-18-8	
1	2	2051-60-7	34
2	3	2051-61-8	16.5, 34
3	4	2051-62-9	77.7
	Dichlorobiphenyl	25512-42-9	
4	2,2’	13029-08-8	60.5
5	2,3	16605-91-7	27.7-28.2
6	2,3’	25569-80-6	oil
7	2,4	33284-50-3	24.1-24.4
8	2,4’	34883-43-7	46
9	2,5	34883-39-1	22-23?
10	2,6	33146-45-1	35-36
11	3,3’	2050-67-1	29
12	3,4	2974-92-7	49-50
13	3,4’	2974-90-5	oil
14	3,5	34883-41-5	31-32
15	4,4’	2050-68-2	148-149
	Trichlorobiphenyl	25323-68-6	
16	2,2’,3	38444-78-9	28.1-28.8
17	2,2’,4	37680-66-3	
18	2,2’,5	37680-65-2	43-44, 45
19	2,2’,6	38444-73-4	
20	2,3,3’	38444-84-7	44.5

Appendix Table 1. Nomenclature of Polychlorinated Biphenyls

IUPAC No.	Compound	CAS No.	Melting Pt.
21	2,3,4	55702-46-0	101-102
22	2,3,4'	38444-85-8	73-73.2
23	2,3,5	55720-44-0	41
24	2,3,6	58702-45-9	50
25	2,3',4	55712-37-3	
26	2,3',5	38444-81-4	40-40.5
27	2,3',6	38444-76-7	
28	2,4,4'	7012-37-5	57-58
29	2,4,5	15862-07-4	78-79
30	2,4,6	35693-92-6	62.5
31	2,4',5	16606-02-3	67
32	2,4',6	38444-77-8	
33	2',3,4	38444-86-9	60
34	2',3,5	37680-68-5	58
35	3,3',4	37680-69-6	68.5
36	3,3',5	38444-87-0	
37	3,4,4'	38444-90-5	86.8-87.8
38	3,4,5	53555-66-1	
39	3,4',5	38444-88-1	88
	Tetrachlorobiphenyl	26914-33-0	
40	2,2',3,3'	38444-93-8	119.5-121.5
41	2,2',3,4	52663-59-9	
42	2,2' 3,4'	36559-22-5	68-70
43	2,2',3,5	70362-46-8	

Appendix Table 1. Nomenclature of Polychlorinated Biphenyls

IUPAC No.	Compound	CAS No.	Melting Pt.
44	2,2',3,5'	41464-39-5	46.5-47, 49
45	2,2',3,6	70362-45-7	
46	2,2' 3,6'	41464-47-5	125.5-127
47	2,2' 4,4'	2437-79-8	32, 41, 83
48	2 2',4,5	70362-47-9	
49	2 2',4,5'	41464-40-8	64-66
50	2,2',4,6	62796-65-8	45
51	2,2',4,6'	65194-04-7	
52	2,2',5,5'	35693-99-3	87-89
53	2,2',5,6'	41464-41-9	103-104.5
54	2,2',6,6'	15968-05-5	101, 198
55	2,3,3',4	74338-24-2	101, 198
56	2,3,3',4'	41464-43-1	96-97
57	2,3,3',5	70424-67-8	
58	2,3,3',5'	41464-49-7	127.5-129
59	2,3,3',6	74472-33-6	
60	2,3,4,4'	33025-41-1	142
61	2,3,4,5	33284-53-6	92-92.5
62	2,3,4,6	54230-23-7	
63	2,3,4',5	74472-35-8	104
64	2,3,4',6	52663-58-8	
65	2,3,5,6	33284-54-7	79
66	2,3',4,4'	32598-10-0	124, 126.5
67	2,3',4,5	73575-53-8	

Appendix Table 1. Nomenclature of Polychlorinated Biphenyls

IUPAC No.	Compound	CAS No.	Melting Pt.
68	2,3',4,5'	73575-52-7	
69	2,3',4,6	60233-24-1	47
70	2,3',4',5	32598-11-1	104
71	2,3',4',6	41464-46-4	
72	2,3',5,5'	41464-42-0	105.5-106.5
73	2,3',5',6	74338-23-1	
74	2,4,4',5	32690-93-0	125
75	2,4,4',6	32598-12-2	65.5
76	2',3,4,5	70362-48-0	
77	3,3',4,4'	32598-13-3	173, 178
78	3,3',4,5	70362-49-1	
79	3,3',4,5'	41464-48-6	119-120
80	3,3',5,5'	33284-52-5	164
81	3,4,4',5	70362-50-4	158
	Pentachlorobiphenyl	25429-29-2	
82	2,2',3,3',4	52663-62-4	119-120.5
83	2,2',3,3',5	60145-20-2	
84	2,2',3,3',6	52663-60-2	
85	2,2',3,4,4'	65510-45-4	
86	2,2',3,4,5	55312-69-1	112
87	2,2',3,4,5'	38380-02-8	111.5-113
88	2,2',3,4,6	55215-17-3	63.5
89	2,2',3,4,6'	73575-57-2	
90	2,2',3,4',5	68194-07-0	

Appendix Table 1. Nomenclature of Polychlorinated Biphenyls

IUPAC No.	Compound	CAS No.	Melting Pt.
91	2,2',3,4',6	58194-05-8	
92	2,2',3,5,5'	52663-61-3	
93	2,2',3,5,6	73575-56-1	
94	2,2',3,5,6'	73575-55-0	
95	2,2',3,5',6	38379-99-6	98.5-100
96	2,2',3,6,6'	73575-54-9	
97	2,2',3',4,5	41464-51-1	81-82
98	2,2',3',4,6	60233-25-2	
99	2,2',4,4',5	38380-01-7	
100	2,2',4,4',6	39485-83-1	34
101	2,2',4,5,5'	37680-73-2	76.5-77.5
102	2,2',4,5,6'	68194-06-9	
103	2,2',4,5',6	60145-21-3	67.5
104	2,2',4,6,6'	56558-16-8	85.5
105	2,3,3',4,4'	32598-14-4	101-105, 115
106	2,3,3',4,5	70424-69-0	
107 (109)	2,3,3',4',5	70424-68-9	
108 (107)	2,3,3',4,5'	70362-41-3	
109 (108)	2,3,3',4,6	74472-35-8	
110	2,3,3',4',6	38380-03-9	oil
111	2,3,3',5,5	39635-32-0	
112	2,3,3',5,6	74472-36-9	91.5
113	2,3,3',5',6	68194-10-5	
114	2 3,4,4',5	74472-37-0	98-99

Appendix Table 1. Nomenclature of Polychlorinated Biphenyls

IUPAC No.	Compound	CAS No.	Melting Pt.
115	2,3,4,4',6	74472-38-1	54
116	2,3,4,5,6	18259-05-7	123
117	2,3,4',5,6	68194-11-6	
118	2,3',4,4',5	31508-00-6	105-107, 109
119	2,3'4,4',6	56558-17-9	77.5
120	2,3'4,5,5'	68194-12-7	
121	2,3',4,5',6	56558-18-0	94
122	2',3,3',4,5	76842-07-4	118
123	2',3,4,4',5	65510-44-3	
124	2',3,4,5,5'	70424-70-3	
125	2',3,4,5,6'	74472-39-2	
126	3,3',4,4',5	57465-28-8	154
127	3,3',4,5,5'	39635-33-1	
	Hexachlorobiphenyl	26601-64-9	
128	2,2',3,3',4,4'	38380-07-3	145.5-146.6
129	2,2',3,3',4,5	55215-18-4	101.5
130	2,2',3,3',4,5'	52663-66-8	
131	2,2',3,3',4,6	61798-70-7	
132	2,2',3,3',4,6'	38380-05-1	
133	2,2',3,3',5,5'	35694-04-3	128-129
134	2,2',3,3',5,6	52704-70-8	
135	2,2',3,3',5,6'	52744-13-5	
136	2,2',3,3',6,6'	38411-22-2	114-114.5
137	2,2',3,4,4',5	35694-06-5	77-78

Appendix Table 1. Nomenclature of Polychlorinated Biphenyls

IUPAC No.	Compound	CAS No.	Melting Pt.
138	2,2',3,4,4',5'	35065-28-2	78.5-80
139	2,2',3,4,4',6	56030-56-9	73
140	2,2',3,4,4',6'	59291-64-4	69.5-71
141	2,2',3,4,5,5'	52712-04-6	85
142	2,2',3,4,5,6	41411-61-4	134-137
143	2,2',3,4,5,6'	68194-15-0	89
144	2,2',3,4,5',6	68194-14-9	
145	2,2',3,4,6,6'	74472-40-5	
146	2,2',3,4',5,5'	51908-16-8	
147	2,2',3,4',5,6	68194-13-8	
148	2,2',3,4',5,6'	74472-42-7	
149	2,2',3,4',5',6	38380-04-0	oil
150	2,2',3,4',6,6'	68194-08-1	
151	2,2',3,5,5',6	52663-63-5	100-101
152	2,2',3,5,6,6'	68194-09-2	
153	2,2',4,4',5,5'	35065-27-1	103-104
154	2,2',4,4',5,6'	60145-22-4	oil, 68.5
155	2,2',4,4',6,6'	33979-03-2	112.5
156	2,3,3',4,4',5	38380-08-4	128
157	2,3,3',4,4',5'	69782-90-7	
158	2,3,3',4,4',6	74472-42-7	107
159	2,3,3',4,5,5'	39635-35-3	132
160	2,3,3',4,5,6	41411-62-5	97-100
161	2,3,3',4,5',6	74472-43-8	

Appendix Table 1. Nomenclature of Polychlorinated Biphenyls

IUPAC No.	Compound	CAS No.	Melting Pt.
162	2,3,3',4',5,5'	39635-34-2	
163	2,3,3',4',5,6	74472-44-9	
164	2,3,3',4',5',6	74472-45-0	
165	2,3,3',5,5',6	74472-46-1	
166	2,3,4,4',5,6	41411-63-6	160-165
167	2,3',4,4',5,5'	52663-72-6	125
168	2,3',4,4',5',6	59291-65-5	110-111
169	3,3',4,4',5,5'	32774-16-6	201-202
	Heptachlorobiphenyl	28655-71-2	
170	2,2',3,3',4,4',5	35065-30-6	134.5-135.5
171	2,2',3,3',4,4',6	52663-71-5	117
172	2,2',3,3',4,5,5'	52663-74-8	
173	2,2',3,3',4,5,6	68194-16-1	
174	2,2',3,3'4,5,6'	38411-25-5	130.5-130.7
175	2,2',3,3'4,5',6	40186-70-7	
176	2,2',3,3',4,6,6'	52663-65-7	
177	2,2',3,3',4',5,6	52663-70-4	
178	2,2',3,3',5,5',6	52663-67-9	
179	2,2',3,3',5,6,6'	52663-64-6	
180	2,2',3,4,4',5,5'	35065-29-3	109-110
181	2,2',3,4,4',5,6	74472-47-2	122
182	2,2',3,4,4',5,6'	60145-23-5	152-153
183	2,2',3,4,4',5',6	52663-69-1	83
184	2,2',3,4,4',6,6'	74472-48-3	

Appendix Table 1. Nomenclature of Polychlorinated Biphenyls

IUPAC No.	Compound	CAS No.	Melting Pt.
185	2,2',3,4,5,5',6	52712-05-7	147-150
186	2,2',3,4,5,6,6'	74472-49-4	117
187	2,2',3,4',5,5',6	52663-68-0	104.5
188	2,2',3,4',5,6,6'	74487-85-7	132
189	2,3,3',4,4',5,5'	39635-31-9	162-163
190	2,3,3',4,4',5,6	41411-64-7	116-118
191	2,3,3',4,4',5',6	74472-50-7	
192	2,3,3',4,5,5',6	74472-51-8	
193	2,3,3',4',5,5',6	69782-91-8	
	Octachlorobiphenyl	31472-83-0	
194	2,2',3,3',4,4',5,5'	35694-08-7	156-157
195	2,2',3,3',4,4',5,6	52663-78-2	168
196	2,2',3,3',4,4',5,6'	42740-50-1	
197	2,2',3,3',4,4',6,6'	33091-17-7	132
198	2,2',3,3',4,5,5',6	68194-17-2	168
199 (201)	2,2',3,3',4,5,5',6'	52663-75-9	189
200 (199)	2,2',3,3',4,5,6,6'	52663-73-7	
201 (200)	2,2',3,3',4,5',6,6'	40186-71-8	
202	2,2',3,3',5,5',6,6'	2136-99-4	161
203	2,2',3,4,4',5,5',6	52663-76-0	
204	2,2',3,4,4' 5,6,6'	74472-52-9	178.5
205	2,3,3',4,4' 5,5',6	74472-53-0	
	Nonachlorobiphenyl	53742-07-7	
206	2,2',3,3',4,4',5,5',6	40186-72-9	204.5-206.5

Appendix Table 1. Nomenclature of Polychlorinated Biphenyls

IUPAC No.	Compound	CAS No.	Melting Pt.
207	2,2',3,3',4,4',5,6,6'	52663-79-3	
208	2,2',3,3',4,5,5',6,6'	52663-77-1	
	Decachlorobiphenyl	2051-24-3	
209	2,2',3,3',4,4',5,5',6,6'	2051-24-3	305-306

Appendix Table 2.

PCB Congener Weight%s in Aroclors of 5 Types

IUPAC Structure No.	Ring Cl Pos.	Lot A2 A1016	Mean 3 Lots A1242	Lot A3.5 A1248	Lot G3.5 A1248	Lot A4 A1254	Lot G4 A1254	Mean 3 Lots A1260
1	2	.52	.54	.05	.02	.02		.02
2	3	.02	.03					
3	4	.15	.18	.01				
4	2-2	3.62	3.08	.32	.04	.02	.06	.02
5	23	.17	.14	.00				
6	2-3	1.64	1.43	.13	.00	.01	.02	.01
7	24	.29	.26	.02				
8	2-4	8.29	7.05	.81	.26	.05	.13	.04
9	25	.58	.50	.04				
10	26	.23	.20					
11	3-3							
12	34	.07	.06					
13	3-4	.24	.22	.02				
14	35							
15	4-4	2.40	2.10	.22	.06	.01	.03	.01
16	23-2	3.88	3.14	1.04	.71	.02	.09	.01
17	24-2	3.98	3.13	1.05	.93	.02	.08	.02
18	25-2	10.86	8.53	4.29	3.29	.08	.25	.05
19	26-2	.99	.80	.22	.14			
20	23-3	.88	.72	.14	.08			
21	234	NM	NM					
22	23-4	3.50	2.84	1.33	1.38	.02	.04	.01
23	235	.01	.01		.00			
24	236	.16	.13	.01				
25	24-3	.72	.59	.11	.04			
26	25-3	1.57	1.28	.40	.23		.03	
27	26-3	.51	.41	.12	.07			
28	24-4	8.50	6.86	3.59	5.57	.06	.19	.03
29	245	.10	.08	.00	.01			
30	246	.00						
31	25-4	9.32	7.34	5.07	5.47	.11	.28	.04
32	26-4	2.37	1.90	.88	.93	.01	.05	.01
33	34-2	6.21	5.01	2.23	2.21	.05	.16	.03
34	35-2	.03	.02	.00	.00			
35	34-3	.05	.08	.00				
36	35-3							
37	34-4	1.02	2.03	.79	.95	.01	.07	.01
38	345							
39	35-4							
40	23-23	.58	.76	1.13	.92	.15	.12	

Appendix Table 2.

PCB Congener Weight%s in Aroclors of 5 Types

IUPAC No.	Structure Ring Cl Pos.	Lot A2 A1016	Mean 3 Lots A1242	Lot A3.5 A1248	Lot G3.5 A1248	Lot A4 A1254	Lot G4 A1254	Mean 3 Lots A1260
41	234-2	.76	.68	.77	.75	.02	.01	
42	23-24	1.59	1.19	1.67	1.79	.09	.15	.01
43	235-2	.28	.18	.30	.19			
44	23-25	4.47	3.55	6.31	5.09	.67	2.31	.03
45	236-2	1.23	.89	1.09	.91	.02	.05	
46	23-26	.49	.36	.47	.39			
47	24-24	1.26	.93	1.49	2.41	.07	.14	
48	245-2	1.61	1.18	1.66	1.54	.05	.12	
49	24-25	3.35	2.53	4.12	4.17	.26	1.10	.01
50	246-2	.01	.00					
51	24-26	.32	.23	.30	.31			
52	25-25	4.63	3.53	6.93	5.58	.83	5.38	.24
53	25-26	.95	.71	1.05	.88	.04	.12	
54	26-26	.01	.01		.01			
55	234-3		.10	.06	.05			
56	23-34	.07	1.81	3.16	3.19	1.70	.55	.02
57	235-3	.01	.02	.02	.02			
58	23-35							
59	236-3	.41	.32	.37	.23	.01	.02	
60	234-4	.04	1.18	1.85	2.67	.95	.18	.04
61	2345							
62	2346							
63	235-4	.06	.12	.17	.19	.07	.02	
64	236-4	1.87	1.70	3.01	3.32	.36	.59	.01
65	2356							
66	24-34	.39	3.39	5.84	7.22	3.56	1.01	.02
67	245-3	.06	.16	.13	.10	.01		
68	24-35							
69	246-3	.00						
70	25-34	.59	3.73	7.28	7.39	6.83	3.49	.04
71	26-34	1.16	1.03	1.67	1.86	.11	.15	.01
72	25-35	.00	.01	.02	.01			
73	26-35	.00	.00					
74	245-4	.33	1.81	3.14	4.67	2.19	.84	.05
75	246-4	.06	.04	.08	.08			
76	345-2		.08	.13	.13	.03	.02	
77	34-34		.31	.41	.52	.20	.03	
78	345-3							
79	34-35							
80	35-35							

Appendix Table 2.

PCB Congener Weight%s in Aroclors of 5 Types

IUPAC No.	Structure Ring Cl Pos.	Lot A2 A1016	Mean 3 Lots A1242	Lot A3.5 A1248	Lot G3.5 A1248	Lot A4 A1254	Lot G4 A1254	Mean 3 Lots A1260
81	345-4		.01	.01	.02	.00		
82	234-23		.26	.81	.62	1.53	1.11	
83	235-23		.11	.26	.20	.56	.48	.01
84	236-23	.05	.41	1.26	.91	1.58	2.32	.11
85	234-24		.31	.98	1.14	2.49	1.28	.01
86	2345-2		.03	.11	.09	.10	.06	
87	234-25		.46	1.45	1.11	3.41	3.99	.41
88	2346-2		.00	.02	.02			
89	234-26		.09	.20	.17	.11	.09	
90	235-24			NM	NM	NM	NM	
91	236-24	.06	.21	.63	.56	.53	.93	.01
92	235-25		.09	.38	.25	.57	1.29	.30
93	2356-2		.00	.04	.03			
94	235-26		.01	.03	.02	.01	.02	
95	236-25	.31	.61	1.96	1.43	1.84	6.25	2.45
96	236-26	.04	.03	.08	.06	.01	.04	
97	245-23		.38	1.22	.97	2.78	2.62	.09
98	246-23							
99	245-24	.01	.46	1.47	1.81	4.53	3.02	.04
100	246-24							
101	245-25	.04	.69	2.22	1.89	5.49	8.02	3.13
102	245-26	.04	.07	.19	.17	.09	.15	
103	246-25			.02	.01		.03	
104	246-26							
105	234-34	.00	.47	1.60	1.45	7.37	2.99	.22
106	2345-3							
107	234-35							
108	2346-3							
109	235-34		.06	.18	.13	.78	.37	.01
110	236-34		.83	2.97	2.55	8.42	9.29	1.33
111	235-35							
112	2356-3							
113	236-35					.01		
114	2345-4		.04	.12	.12	.50	.18	
115	2346-4		.04	.11	.11	.37	.20	
116	23456							
117	2356-4		.03	.09	.10	.19	.23	
118	245-34		.66	2.29	2.35	13.59	7.35	.48
119	246-34			.06	.06	.12	.08	
120	245-35							

Appendix Table 2.

PCB Congener Weight%s in Aroclors of 5 Types

IUPAC No.	Structure Ring Cl Pos.	Lot A2 A1016	Mean 3 Lots A1242	Lot A3.5 A1248	Lot G3.5 A1248	Lot A4 A1254	Lot G4 A1254	Mean 3 Lots A1260
121	246-35							
122	345-23		.01	.06	.05	.25	.10	
123	345-24		.03	.07	.08	.32	.15	
124	345-25		.03	.10	.07	.47	.29	.01
125	345-26		.02	.04	.03	.03	.02	
126	345-34			.00	.00	.02	.00	
127	345-35							
128	234-234		.02	.12	.08	1.71	1.42	.53
129	2345-23			.02		.39	.38	.14
130	234-235			.04	.01	.50	.60	.22
131	2346-23					.14	.19	.07
132	234-236		.04	.15	.14	1.50	2.29	2.90
133	235-235						.11	.07
134	2356-23				.01	.20	.37	.34
135	235-236			.04	.04	.28	.61	1.08
136	236-236			.05	.06	.24	.70	1.46
137	2345-24			.03	.02	.52	.42	.02
138	234-245		.10	.38	.41	5.95	5.80	6.54
139	2346-24					.14	.15	
140	234-246							
141	2345-25		.01	.07	.09	.69	.98	2.62
142	23456-2							
143	2345-26							
144	2346-25				.01	.12	.24	.61
145	2346-26							
146	235-245			.04	.05	.45	.67	1.15
147	2356-24					.02	.10	
148	235-246							
149	236-245		.06	.24	.33	1.82	3.65	8.75
150	236-246							
151	2356-25			.04	.08	.22	.69	3.04
152	2356-26							
153	245-245		.06	.23	.43	3.29	3.77	9.39
154	245-246					.02	.04	
155	246-246							
156	2345-34		.01	.06	.04	1.13	.82	.52
157	234-345			.01	.00	.30	.19	.02
158	2346-34		.01	.04	.04	.90	.81	.58
159	2345-35							
160	23456-3							

Appendix Table 2.

PCB Congener Weight%s in Aroclors of 5 Types

IUPAC No.	Structure Ring Cl Pos.	Lot A2 A1016	Mean 3 Lots A1242	Lot A3.5 A1248	Lot G3.5 A1248	Lot A4 A1254	Lot G4 A1254	Mean 3 Lots A1260
161	2346-35							
162	235-345							
163	2356-34		.01	.06	.08	.70	1.03	2.42
164	236-345			.02	.03	.31	.40	.69
165	2356-35							
166	23456-4					.05	.05	
167	245-345			.01	.01	.35	.27	.19
168	246-345							
169	345-345							
170	2345-234				.08	.35	.52	4.11
171	2346-234					.08	.14	1.11
172	2345-235					.03	.07	.70
173	23456-23							.10
174	2345-236				.08	.14	.34	4.96
175	2346-235							.17
176	2346-236					.01	.04	.59
177	2356-234				.03	.08	.20	2.57
178	2356-235						.03	.83
179	2356-236				.02	.02	.10	2.03
180	2345-245			.02	.21	.42	.67	11.38
181	23456-24							.01
182	2345-246							
183	2346-245				.06	.09	.18	2.41
184	2346-246							
185	23456-25							.55
186	23456-26							
187	2356-245				.09	.09	.25	5.40
188	2356-246							
189	2345-345					.01	.01	.10
190	23456-34					.05	.07	.82
191	2346-345							.17
192	23456-35							
193	2356-345						.03	.53
194	2345-2345						.01	2.07
195	23456-234							.84
196	2345-2346							1.09
197	2346-2346							.07
198	23456-235							.10
199	2345-2356						.01	1.78
200	23456-236							.25

Appendix Table 2.

PCB Congener Weight%s in Aroclors of 5 Types

IUPAC No.	Structure Ring Cl Pos.	Lot A2 A1016	Mean 3 Lots A1242	Lot A3.5 A1248	Lot G3.5 A1248	Lot A4 A1254	Lot G4 A1254	Mean 3 Lots A1260
201	2346-2356							.24
202	2356-2356							.33
203	23456-245						.02	1.40
204	23456-246							
205	23456-345							.10
206	23456-2345					.03	.03	.53
207	23456-2346							.05
208	23456-2356					.01	.01	.13
209	23456-23456							NM
Sum of Wt%s =		100.0	100.0	100.2	100.2	100.2	100.4	100.3

Homolog % Dist.	A2 A1016	Mean A1242	A3.5 A1248	G3.5 A1248	A4 A1254	G4 A1254	Mean A1260
mono-Cl	.70	.75	.07	.02	.02	.00	.02
di-Cl	17.53	15.04	1.55	.36	.09	.24	.08
tri-Cl	54.67	44.91	21.27	22.02	.39	1.26	.21
tetra-Cl	22.07	20.16	32.77	31.12	4.86	10.25	.35
penta-Cl	5.07	18.85	42.92	44.09	71.44	59.12	8.74
hexa-Cl	.00	.31	1.64	1.97	21.97	26.76	43.35
hepta-Cl	.00	.00	.02	.57	1.36	2.66	38.54
octa-Cl	.00	.00	.00	.00	.00	.04	8.27
nona-Cl	.00	.00	.00	.00	.04	.04	.70

N.B. IUPAC#s 107, 108, & 109 here correspond to BZ #s 108, 109 & 107.
IUPAC#s 199, 200, & 201 here correspond to BZ #s 201, 199 & 200.
NM indicates congener not measured but present at trace level.

Underlined PCB # indicates present at > 1.0 Wt% in some Aroclor.
Italicized PCB# indicates not present at > 0.10 Wt% in any Aroclor; these "non-Aroclor congeners" generally have major imbalance in no. of Cls/ring and/or less probable 3-, 35- or 246- substituted rings

** Lot A4 Aroclor 1254 from abnormal late production (1974-1977) made by chlorinating A1016 distillate residue and contains elevated percentages of non- and mono-ortho chlorinated congeners relative to "normal" G4 Lot

Data condensed from: Frame, G. M., Cochran, J. W., and Boewadt, S.S., *J. High Res. Chromatogr.*, Vol. 19, pp 657-668 (1996)

Appendix Table 3. Some properties of chlorobiphenyls.

PCB[a]	Log K_{ow}		Henry's Law Constant[d]	Gibbs Free Energies of Formation[e]	
	Ave[b]	Single[c]		Gas	Liquid
1	4.3		3.526	262.7	234
2	4.6		3.544	253.8	222.7
3	4.5		3.562	255	223.6
4	4.9		3.483	237.7	207
5	5		3.622	245.7	212.5
6			3.486	235.9	202.9
7	5		3.424	238.4	205.5
8	5.1		3.518	238.4	205.1
9	5.1		3.49	238.1	205.5
10	5		3.373	241.2	210.2
11	5.3		3.537	227.4	192.1
12	5.3		3.631	235.7	199.9
13			3.595	228.3	192.7
14	5.4		3.376	229.6	194.8
15	5.3		3.649	231.3	195.4
16	5.6		3.6	220.2	185
17			3.428	213.1	178.3
18	5.6		3.495	213.1	178.6
19			3.355	213.7	180.7
20	5.5		3.663	219.7	182.2
21	5.8		3.644	231	193.5
22			3.719	221.4	183.6
23			3.497	221.4	184.9

Appendix Table 3. Some properties of chlorobiphenyls.

PCB[a]	Log K_{ow}		Henry's Law Constant[d]	Gibbs Free Energies of Formation[e]	
	Ave[b]	Single[c]		Gas	Liquid
24			3.507	222.5	187.9
25			3.5	212.6	175.5
26	5.5		3.526	212.6	175.8
27			3.393	215.4	180.1
28	5.8	5.67	3.544	214.3	176.9
29	5.6		3.529	221.4	184.7
30	5.5		3.242	217.1	182.7
31	5.7	5.67	3.562	214.3	177.2
32			3.407	217.1	181.5
33	5.8		3.62	219.7	182
34			3.375	214.3	177.5
35	5.7		3.745	210.4	170.3
36			3.473	205	165.9
37	5.9		3.818	212.2	171.8
38			3.634	223.5	183.6
39			3.524	206.8	167.4
40	5.6		3.738	206.3	166.7
41			3.612	204.6	165.2
42		5.76	3.592	195.4	156.1
43			3.475	195.4	156.9
44	6	5.75	3.638	195.4	156.4
45			3.45	195.4	158.9
46			3.47	197.2	159.7

Appendix Table 3. Some properties of chlorobiphenyls.

	Log K_{ow}			Gibbs Free Energies of Formation[e]	
PCB[a]	Ave[b]	Single[c]	Henry's Law Constant[d]	Gas	Liquid
47	5.9		3.434	188	149
48			3.519	195.4	156.7
49	6.1	5.85	3.452	186.2	147.6
50	5.8		3.215	188	151.6
51	5.7		3.292	188	150.8
52	6.1	5.84	3.496	188	149.6
53	5.5		3.366	188	151.1
54	5.5		3.242	191.3	156
55			3.739	204.6	162.8
56			3.82	204.6	162.4
57			3.568	195.4	154.6
58			3.602	197.2	155.9
59			3.517	195.4	156.5
60	6.3	6.11	3.816	206.4	164.3
61	5.9		3.623	215.6	174.2
62			3.432	206.4	167.5
63			3.615	197.2	156.1
64		5.95	3.565	197.2	158
65	5.9		3.473	208	169
66	5.8	6.2	3.693	195.4	153.3
67			3.631	195.4	154.4
68			3.424	188	147
69			3.296	188	149.3

Appendix Table 3. Some properties of chlorobiphenyls.

PCB[a]	Log K_{ow} Ave[b]	Log K_{ow} Single[c]	Henry's Law Constant[d]	Gibbs Free Energies of Formation[e] Gas	Gibbs Free Energies of Formation[e] Liquid
70	5.9	6.2	3.694	195.4	153.8
71			3.503	197.2	157.1
72			3.441	188	147.3
73			3.284	189.6	150.5
74		6.2	3.668	197.2	155.9
75	6.2		3.333	189.6	150.6
76			3.622	206.4	164.6
77	6.5	6.05	3.989	196.4	151.6
78			3.787	196.4	152.3
79			3.705	187.2	143.4
80	6.4		3.426	181.4	138.5
81			3.844	198.1	153.6
82			3.835	188.7	144.7
83	6.2		3.674	179.5	136.5
84			3.6	179.5	138.4
85			3.716	179.5	135.9
86	6.2		3.623	188.7	149.3
87	6.5	6.29	3.736	179.5	136.2
88	6.5		3.415	179.5	140.7
89			3.526	181.1	139.5
90			3.531	170.3	127.7
91			3.461	170.3	129.6
92			3.585	170.3	128

Appendix Table 3. Some properties of chlorobiphenyls.

PCB[a]	Log K_{ow}		Henry's Law Constant[d]	Gibbs Free Energies of Formation[e]	
	Ave[b]	Single[c]		Gas	Liquid
93			3.468	181.2	142.3
94			3.407	172	131.2
95	6.1		3.523	170.3	129.9
96			3.387	172	133.2
97		6.29	3.745	179.5	136.3
98			3.407	172	131.2
99	6.6	6.39	3.603	170.3	127.4
100	6.2		3.25	162.8	122.3
101	6.4	6.38	3.61	170.3	127.8
102			3.431	172	131
103	6.4		3.298	162.8	122.6
104	6.1		3.13	164.5	125.9
105	6	6.65	4.003	188.7	142.1
106			3.703	188.7	147
107			3.798	179.5	133.9
108			3.755	181.2	135.6
109			3.55	179.5	138.4
110	6.3	6.48	3.707	179.5	135.8
111			3.574	172	127.4
112			3.574	181.2	139.9
113			3.487	172	129.4
114			3.845	190.4	148.4
115			3.61	181.2	139.8

Appendix Table 3. Some properties of chlorobiphenyls.

PCB[a]	Log K_{ow}		Henry's Law Constant[d]	Gibbs Free Energies of Formation[e]	
	Ave[b]	Single[c]		Gas	Liquid
116	6.3		3.529	201.3	158.7
117			3.618	182.9	141.4
118	6.3	6.74	3.901	179.5	133.7
119			3.508	172	128.6
120			3.61	172	127.2
121			3.253	164.5	122.1
122			3.901	190.4	144.1
123			3.759	181.2	135.3
124			3.768	181.2	135.6
125			3.541	182.9	138.8
126		6.89	4.087	180.5	131.6
127			3.807	173	125.1
128	7		3.984	174.4	126.2
129	7.3	6.73	3.854	172.7	128.8
130			3.817	163.5	116.3
131			3.616	163.5	120.2
132			3.693	163.5	118.2
133			3.691	156	109.8
134	7.3		3.639	165.2	121.8
135			3.571	154.3	110
136	6.7		3.492	156	113.7
137			3.731	163.5	120
138	6.8	6.83	3.886	163.5	116.1

Appendix Table 3. Some properties of chlorobiphenyls.

PCB[a]	Log K_{ow}		Henry's Law Constant[d]	Gibbs Free Energies of Formation[e]	
	Ave[b]	Single[c]		Gas	Liquid
139			3.483	154.3	111.4
140			3.512	156	110.9
141		6.82	3.76	163.5	120.3
142			3.502	174.4	131.8
143			3.531	165.2	123.6
144			3.529	154.3	111.7
145			3.328	156	115
146		6.89	3.727	154.3	107.8
147			3.501	156	113
148			3.367	146.8	102.7
149		6.67	3.625	154.3	109.8
150			3.296	146.8	104.7
151			3.548	156	113.3
152		6.22	3.369	157.8	116.6
153	6.9	6.92	3.783	156	109.4
154			3.418	146.8	102.5
155			3.075	141.1	99.1
156			4.053	126.3	168.4
157			4.073	174.4	123.9
158			3.782	163.5	117.6
159			3.808	165.2	119.8
160			3.67	174.4	129.5
161			3.545	156	111.2

Appendix Table 3. Some properties of chlorobiphenyls.

PCB[a]	Log K_{ow} Ave[b]	Log K_{ow} Single[c]	Henry's Law Constant[d]	Gibbs Free Energies of Formation[e] Gas	Gibbs Free Energies of Formation[e] Liquid
162			3.881	165.2	115.7
163			3.781	165.2	119.2
164			3.754	165.2	117.6
165			3.56	157.8	112.8
166			3.735	176.2	130.9
167			3.959	165.2	115.4
168			3.559	157.8	110.3
169	7.5	7.42	4.186	168	115.1
170	7.1	7.4	4.059	156.9	108.8
171	6.7	7.11	3.763	147.7	100.1
172		7.33	3.924	147.7	100.5
173			3.739	158.5	111.4
174		7.11	3.772	147.7	102.5
175			3.651	138.4	91.9
176			3.527	138.4	93.8
177			3.787	149.3	101.6
178			3.671	140.1	93.4
179			3.56	147.7	102.9
180	7.2	7.36	3.969	140.1	92.7
181			3.638	149.3	102.5
182			3.59	140.1	95.1
183		7.2	3.696	138.4	91.7
184			3.339	130.9	86.4

Appendix Table 3. Some properties of chlorobiphenyls.

PCB[a]	Log K_{ow}		Henry's Law Constant[d]	Gibbs Free Energies of Formation[e]	
	Ave[b]	Single[c]		Gas	Liquid
185	7	7.4	3.669	149.3	102.8
186			3.434	151	106.1
187	7.1	7.2	3.693	140.1	93.1
188			3.353	132.6	88
189			4.177	158.5	108
190			3.95	158.5	108.8
191			3.876	149.3	100.5
192			3.781	151	102.3
193			3.872	151	101
194	7.4	7.8	4.174	142.6	94.6
195		7.56	3.926	142.6	91.3
196			3.884	131.7	84.3
197			3.596	124.3	77.4
198			3.812	133.5	83.1
199			3.644	133.5	85
200		7.27	3.619	124.3	76.6
201		7.62	3.884	133.5	85.8
202	7.1		3.651	127.6	80.4
203		7.65	3.853	133.5	82.8
204			3.463	126	77.7
205			4.059	144.4	90.6
206	7.2	8.09	4.059	126.7	75.4
207	7.5		3.772	117.5	66.8

Appendix Table 3. Some properties of chlorobiphenyls.

PCB[a]	Log K_{ow}		Henry's Law Constant[d]	Gibbs Free Energies of Formation[e]	
	Ave[b]	Single[c]		Gas	Liquid
208	8.2		3.777	119.2	68.4
209	8.3		3.948	114.2	59.7

[a)] See Table 1 for structures.
[b)] From Mackay et al., 1992. If no value from several was selected by the authors, approximate mean values for several determinations and/or calculations are listed.
[c)] From Koslowski et al., 1994.
[d)] From Dunnivant et al., 1992. Negative log of Henry's Law Constants (atm·m^3/mol at 25° C).
[e)] Gibbs Free Energies of Formation at 298.15° K and 100,000 Pa (Holmes et al., 1993).

Appendix Table 4. Estimation of PCB Congener Production Ratios. Composition data from Frame *et al.*, 1996a; ratios are for comparative USA sales from Table 1.2 and 1.3.

PCB	A1221	A1232	A1242	A1248	A1254	A1260	A1262	Relative Prod'n	
Ratio	0.0075	0.0075	0.598	0.088	0.153	0.135	0.011	From Table 1.3	From Table 1.2
Ratio	0.0026	0.0026	0.187	0.227	0.266	0.3	0.015		
1	35.80	15.53	0.54	0.04	0.00	0.02	0.03	0.72	0.26
2	3.81	1.96	0.03	0	0	0	0	0.06	0.02
3	20.44	10.28	0.18	0.01	0.00	0.00	0.01	0.34	0.12
4	6.19	5.35	3.08	0.18	0.06	0.02	0.05	1.96	0.70
5	0.74	0.49	0.14	0.002	0	0.000	0	0.10	0.03
6	3.82	3.01	1.43	0.07	0.02	0.01	0.02	0.92	0.32
7	1.70	1.11	0.26	0.009	0	0	0	0.18	0.06
8	12.34	10.71	7.05	0.53	0.13	0.04	0.11	4.47	1.63
9	1.74	1.27	0.50	0.02	0	0	0.001	0.32	0.11
10	0.80	0.59	0.20	0	0	0	0	0.13	0.04
11	0.16	0	0	0	0	0	0	0.001	0.000
12	0.59	0.35	0.06	0	0	0	0	0.04	0.02
13	1.12	0.72	0.22	0.009	0	0	0	0.14	0.05
14	0.001	0.012	0	0	0	0	0	0.000	0.000
15	4.18	3.21	2.10	0.14	0.03	0.01	0.03	1.33	0.48
16	0.31	1.79	3.14	0.88	0.09	0.01	0.05	1.99	0.84
17	0.34	1.83	3.13	0.99	0.08	0.02	0.05	1.99	0.86
18	0.78	4.86	8.53	3.79	0.25	0.05	0.15	5.53	2.56
19	0.08	0.46	0.80	0.18	0.00	0.00	0.01	0.50	0.20
20	0.07	0.42	0.72	0.11	0	0	0	0.44	0.17
22	0.26	1.62	2.84	1.35	0.04	0.01	0.05	1.84	0.86
23	0	0.006	0.011	0.002	0	0	0	0.01	0.00
24	0.02	0.08	0.13	0.00	0	0	0	0.08	0.03
25	0.09	0.37	0.59	0.07	0	0	0.01	0.37	0.13
26	0.13	0.75	1.28	0.31	0.035	0.004	0.02	0.80	0.33

Appendix Table 4. Estimation of PCB Congener Production Ratios. Composition data from Frame *et al.*, 1996a; ratios are for comparative USA sales from Table 1.2 and 1.3.

PCB	A1221	A1232	A1242	A1248	A1254	A1260	A1262	Relative Prod'n	
27	0.05	0.12	0.41	0.10	0	0	0.004	0.26	0.10
28	0.62	3.91	6.86	4.58	0.19	0.03	0.11	4.59	2.36
29	0.013	0.053	0.081	0.007	0	0	0	0.05	0.02
31	0.60	4.14	7.34	5.27	0.28	0.04	0.12	4.95	2.62
32	0.17	1.07	1.90	0.90	0.05	0.01	0.03	1.24	0.58
33	0.48	2.86	5.01	2.22	0.16	0.03	0.10	3.25	1.51
34	0.000	0.013	0.025	0.004	0	0	0	0.02	0.01
35	0.010	0.051	0.076	0.002	0	0	0	0.05	0.02
37	0.19	1.13	2.03	0.87	0.07	0.01	0.03	1.31	0.60
40	0.04	0.38	0.76	1.03	0.12	0.00	0.01	0.57	0.39
41	0.03	0.35	0.68	0.76	0.01	0.00	0.01	0.48	0.29
42	0.09	0.68	1.19	1.73	0.15	0.01	0.02	0.89	0.63
43	0	0.10	0.18	0.24	0	0	0	0.13	0.08
44	0.21	1.81	3.55	5.70	2.31	0.035	0.079	3.01	2.48
45	0.04	0.46	0.89	1.00	0.05	0.003	0.007	0.63	0.39
46	0.02	0.19	0.36	0.43	0	0.001	0.001	0.25	0.16
47	0.05	0.49	0.93	1.95	0.14	0.003	0.011	0.76	0.62
48	0.06	0.61	1.18	1.60	0.12	0.003	0.011	0.87	0.59
49	0.15	1.36	2.53	4.15	1.10	0.014	0.058	2.07	1.63
50	0	0	0.005	0	0	0	0	0.00	0.00
51	0.01	0.12	0.23	0.31	0	0	0	0.17	0.11
52	0.22	1.84	3.53	6.25	5.38	0.24	0.14	3.54	3.46
53	0.04	0.37	0.71	0.96	0.12	0.00	0.01	0.53	0.37
54	0	0.002	0.008	0.004	0	0	0	0.01	0.00
55	0	0.049	0.099	0.055	0	0	0	0.06	0.03
56	0.12	0.92	1.81	3.18	0.55	0.02	0.03	1.46	1.15
57	0	0.009	0.019	0.023	0	0	0	0.01	0.01
59	0.01	0.18	0.32	0.30	0.02	0.00	0.00	0.22	0.13

Appendix Table 4. Estimation of PCB Congener Production Ratios. Composition data from Frame *et al.*, 1996a; ratios are for comparative USA sales from Table 1.2 and 1.3.

PCB	A1221	A1232	A1242	A1248	A1254	A1260	A1262	Relative Prod'n	
60	0.07	0.60	1.18	2.26	0.18	0.04	0.02	0.95	0.75
63	0.01	0.10	0.12	0.18	0.02	0	0	0.09	0.07
64	0.10	0.87	1.70	3.17	0.59	0.01	0.03	1.40	1.14
66	0.21	1.73	3.39	6.53	1.01	0.02	0.06	2.79	2.26
67	0.01	0.08	0.16	0.11	0	0	0	0.11	0.05
70	0.24	1.90	3.73	7.34	3.49	0.04	0.10	3.45	3.16
71	0.06	0.54	1.03	1.77	0.15	0.01	0.01	0.81	0.60
72	0	0.001	0.012	0.015	0	0	0	0.01	0.01
73	0	0	0.001	0	0	0	0	0.00	0.00
74	0.12	0.92	1.81	3.91	0.84	0.05	0.05	1.58	1.38
75	0	0.023	0.042	0.08	0	0	0	0.03	0.02
76	0.01	0	0.08	0.13	0.02	0	0	0.06	0.05
77	0.01	0.16	0.31	0.47	0.03	0	0	0.23	0.16
81	0	0.004	0.011	0.017	0	0	0	0.01	0.01
82	0.00	0.12	0.26	0.72	1.11	0	0	0.39	0.49
83	0	0.04	0.11	0.23	0.48	0.01	0.00	0.16	0.20
84	0.02	0.19	0.41	1.09	2.32	0.11	0.04	0.71	0.95
85	0.03	0.17	0	1.06	1.28	0.01	0.02	0.29	0.55
86	0	0.013	0	0.098	0.060	0	0	0.02	0.04
87	0.04	0.22	0.46	1.28	3.99	0.41	0.11	1.06	1.53
88	0	0	0	0.02	0	0	0	0.00	0.00
89	0	0.05	0.08	0.185	0.09	0	0	0.08	0.08
91	0	0.10	0.21	0.59	0.93	0.01	0.01	0.32	0.41
92	0.02	0.05	0.09	0.31	1.29	0.30	0.08	0.32	0.51
93	0	0	0	0.032	0	0	0	0.00	0.01
94	0	0	0.01	0.027	0.024	0	0	0.01	0.01
95	0.05	0.30	0.61	1.70	6.25	2.45	0.93	1.82	2.87
96	0	0.01	0.02	0.07	0.04	0	0	0.02	0.03

Appendix Table 4. Estimation of PCB Congener Production Ratios. Composition data from Frame *et al.*, 1996a; ratios are for comparative USA sales from Table 1.2 and 1.3.

PCB	A1221	A1232	A1242	A1248	A1254	A1260	A1262	Relative Prod'n	
97	0.03	0.18	0.38	1.10	2.62	0.09	0.04	0.74	1.02
99	0.04	0.21	0.46	1.64	3.02	0.04	0.05	0.89	1.24
101	0.07	0.33	0.69	2.05	8.02	3.13	1.13	2.26	3.63
102	0	0.03	0.07	0.18	0.15	0	0	0.08	0.09
103	0	0	0	0.01	0.03	0	0	0.00	0.01
105	0.05	0.21	0.47	1.53	2.99	0.22	0.09	0.91	1.26
109	0	0.03	0.06	0.16	0.37	0.01	0.01	0.11	0.14
110	0.05	0.38	0.83	2.76	9.29	1.33	0.39	2.35	3.59
114	0	0.02	0.04	0.12	0.18	0.001	0	0.06	0.08
115	0	0.01	0.04	0.11	0.20	0.001	0	0.07	0.08
117	0	0.01	0.03	0.10	0.23	0	0	0.06	0.09
118	0.08	0.29	0.66	2.32	7.35	0.48	0.15	1.80	2.69
119	0	0	0	0.06	0.08	0	0	0.02	0.03
122	0	0	0.01	0.06	0.10	0	0	0.03	0.04
123	0	0	0.03	0.08	0.15	0	0	0.05	0.06
124	0	0.00	0.03	0.08	0.29	0.007	0.003	0.07	0.10
125	0	0.01	0.02	0.04	0.02	0	0	0.02	0.02
126	0	0	0	0.003	0.002	0	0	0.001	0.001
128	0	0	0.02	0.10	1.42	0.53	0.19	0.31	0.56
129	0	0	0.00	0.01	0.38	0.14	0.03	0.08	0.14
130	0	0	0	0.03	0.60	0.22	0.04	0.12	0.23
131	0	0	0	0	0.19	0.07	0	0.04	0.07
132	0	0.02	0.04	0.15	2.29	2.90	1.21	0.79	1.53
133	0	0	0	0	0.11	0.07	0.04	0.03	0.05
134	0	0	0	0.01	0.37	0.34	0.13	0.11	0.21
135	0	0	0	0.04	0.61	1.08	0.66	0.25	0.50
136	0	0.003	0.001	0.05	0.70	1.46	1.00	0.32	0.65
137	0	0	0	0.03	0.42	0.02	0.01	0.07	0.12

Appendix Table 4. Estimation of PCB Congener Production Ratios. Composition data from Frame *et al.*, 1996a; ratios are for comparative USA sales from Table 1.2 and 1.3.

PCB	A1221	A1232	A1242	A1248	A1254	A1260	A1262	Relative Prod'n	
138	0	0.06	0.10	0.39	5.80	6.54	2.74	1.89	3.64
139	0	0	0	0	0.15	0	0	0.02	0.04
141	0	0	0.01	0.08	0.98	2.62	1.66	0.53	1.09
144	0	0	0	0.01	0.24	0.61	0.41	0.12	0.25
146	0	0	0	0.04	0.67	1.15	0.59	0.27	0.54
147	0	0	0	0	0.102	0	0	0.02	0.03
149	0	0.05	0.06	0.29	3.65	8.75	6.40	1.86	3.76
151	0	0.01	0.00	0.06	0.69	3.04	2.97	0.55	1.15
153	0.0015	0.05	0.06	0.33	3.77	9.39	7.10	1.98	4.00
154	0	0	0	0	0.04	0	0	0.01	0.01
156	0	0	0.01	0.05	0.82	0.52	0.16	0.20	0.39
157	0	0	0	0.01	0.19	0.02	0.00	0.03	0.06
158	0	0	0.011	0.04	0.81	0.58	0.20	0.22	0.40
163	0	0.015	0.010	0.07	1.03	2.42	1.52	0.51	1.04
164	0	0	0.001	0.02	0.40	0.69	0.27	0.16	0.32
166	0	0	0	0	0.05	0	0	0.01	0.01
167	0	0	0	0.01	0.27	0.19	0.03	0.07	0.13
170	0	0	0	0.04	0.52	4.11	3.26	0.67	1.43
171	0	0	0	0	0.14	1.11	0.87	0.18	0.38
172	0	0	0	0	0.07	0.70	0.62	0.11	0.24
173	0	0	0	0	0	0.10	0.04	0.01	0.03
174	0	0	0	0.04	0.34	4.96	6.33	0.79	1.68
175	0	0	0	0	0	0.17	0.18	0.03	0.05
176	0	0	0	0	0.04	0.59	0.70	0.09	0.20
177	0	0	0	0.02	0.20	2.57	2.77	0.41	0.87
178	0	0	0	0	0.03	0.83	1.21	0.13	0.28
179	0	0	0	0.01	0.10	2.03	3.33	0.33	0.69
180	0	0.02	0.00	0.11	0.67	11.38	14.13	1.80	3.83

Appendix Table 4. Estimation of PCB Congener Production Ratios. Composition data from Frame *et al.*, 1996a; ratios are for comparative USA sales from Table 1.2 and 1.3.

PCB	A1221	A1232	A1242	A1248	A1254	A1260	A1262	Relative Prod'n	
181	0	0	0	0	0	0.01	0	0.0011	0.0024
183	0	0	0	0.03	0.18	2.41	2.88	0.38	0.82
185	0	0	0	0	0	0.55	0.87	0.08	0.18
187	0	0.01	0	0.04	0.25	5.40	9.15	0.87	1.83
189	0	0	0	0	0.01	0.10	0.04	0.02	0.03
190	0	0	0	0	0.07	0.82	0.76	0.13	0.28
191	0	0	0	0	0	0.17	0.13	0.02	0.05
193	0	0	0	0	0.03	0.53	0.66	0.08	0.18
194	0	0	0	0	0.01	2.07	4.06	0.32	0.69
195	0	0	0	0	0	0.84	1.43	0.13	0.27
196	0	0	0	0	0	1.09	2.26	0.17	0.36
197	0	0	0	0	0	0.07	0.13	0.01	0.02
198	0	0	0	0	0	0.10	0.23	0.02	0.03
199	0	0	0	0	0.01	1.78	4.74	0.29	0.61
200	0	0	0	0	0	0.25	0.65	0.04	0.09
201	0	0	0	0	0	0.24	0.62	0.04	0.08
202	0	0	0	0	0	0.33	1.08	0.06	0.12
203	0	0	0	0	0.02	1.40	4.24	0.24	0.49
205	0	0	0	0	0	0.10	0.17	0.02	0.03
206	0	0	0	0	0.03	0.53	1.26	0.09	0.19
207	0	0	0	0	0	0.05	0.17	0.01	0.02
208	0	0	0	0	0.01	0.13	0.28	0.02	0.05

Appendix Table 5. Comparison of human residues reported by sample authors without standards available. A number of structures were considered in the original analytical profile by Fait et al., but these have since been determined to be rare by the authors or others and are not listed.[a] Changes in the published designations provided by the analyst and/or based on new information are indicated by an asterisk. *Note that these are still "best guesses."*

B & Z No.[a]	Fat (ng/g)[b]			Milk (%)[c]	Fat (μg/g)[d]	
	Peak	Worker	General		Conc.	Peak
15	16				0.08	1A
18	11					
28 + 33	22	0-22[b]	5.9	2.4		
(28+) 31	21			0.75	9	1
40,41,75	29					
41 w 40	29				0.3	5A
44	26	1.7	0.95	+	0.5	5
47 + 48	28*				0.6	4
49	27*				0.3	3
52 + 43	25*	6.8	3.4		0.5	2
60	43	7.1	1.2	1.8	0.9	13
66	40	14	4.3	+	2	8
70	38	0-6.4[b]	1.8	see 74	0.4	7
74	39	34	2	(6.4)*	7	6
82 w 87	55*			see 110		
83	46				+	15
84,91	36	4.5	3.1		+	14
85	50	11	2.6			
87 + 82	55*	270	8.4		0.2	17
92 w 56	41				0.1	10
95	34	17	6		0.2	9

Appendix Table 5. Comparison of human residues reported by sample authors without standards available. A number of structures were considered in the original analytical profile by Fait et al., but these have since been determined to be rare by the authors or others and are not listed.[a] Changes in the published designations provided by the analyst and/or based on new information are indicated by an asterisk. *Note that these are still "best guesses."*

	Fat (ng/g)[b]			Milk	Fat (μg/g)[d]	
B & Z No.[a]	Peak	Worker	General	(%)[c]	Conc.	Peak
97, 109	48	2.2	1		+	16
99	47	60	19	4.2	1	12
101	45	31	6.1	1.0	0.1	11
105 + 146	63	36	13	3.7	1	23
109 w 97	48				0.1	19
110	59*	190	60	(1.7)*	0.1	18
114,175	62	350	110			
118,122	60	144	41	(9.4)*	2	20
119	49	15	4.3			
123,178	59	10				
124	58	4.1	0-1.6[b]			
126-->unid'd[e]	75*	52	7.8			
128 + 187	72	120	23	2.0	0.07	29
130 + 135	53*	43	51		0.1	24
131 + 139	56	61	25			
132	57	23	4.1	+		
133 v low	61	8.9	1.3			
136	44	8				
137,164	66	11	3.2			
138	68	270	84	7.9	1	25
144 v low	54	49	8.1			

Appendix Table 5. Comparison of human residues reported by sample authors without standards available. A number of structures were considered in the original analytical profile by Fait et al., but these have since been determined to be rare by the authors or others and are not listed.[a] Changes in the published designations provided by the analyst and/or based on new information are indicated by an asterisk. *Note that these are still "best guesses."*

B & Z No.[a]	Fat (ng/g)[b]			Milk (%)[c]	Fat (μg/g)[d]	
	Peak	Worker	General		Conc.	Peak
146 w 105	63			2.1	0.1	21
149 + 134	55*	210	75	+		
153 + 141	64	210	73	12.1	1	22
	42	5.6	2			
156 + 157	80	61	25	3.8	0.4	32
158	70	4.2	4.4	0.8		
163 + 179	69	120	43	1.0	0.3	25A*
167	78	15	4		0.1	28
170	84	190	60	2.3	0.2	35
171 + 201	77	41		2.2		
172	81	22	6			
174,183,185	73	54	11	0.8	0.09	30
176	71	0.81	0-0.05[b]	(1.17)*		
177,189,202	74	6.4	3.9		0.07	31
179 w 163	69			see 118		
180 + 193	82	350	110	4.5	0.6	33
183 w 174	73				0.1	27
187 w 128	72				0.2	26
189 w 177	74			(1.6)*		
190	86	30	8.5			
191	83	4.1	0-1.6			

Appendix Table 5. Comparison of human residues reported by sample authors without standards available. A number of structures were considered in the original analytical profile by Fait et al., but these have since been determined to be rare by the authors or others and are not listed.[a] Changes in the published designations provided by the analyst and/or based on new information are indicated by an asterisk. *Note that these are still "best guesses."*

B & Z No.[a]	Fat (ng/g)[b]			Milk (%)[c]	Fat (μg/g)[d]	
	Peak	Worker	General		Conc.	Peak
194	93	46	20			37
195	90	0-71	0-7.0			
196	87	36	13		0.1	36
198,199	85	120	43			
199 w 198	85				0.2	34
201 w 159	77			0.7		
203	88	46	21	1.1		
206	96	18	13			
208	89	4.2	4.4			

[a] Congeners considered in original peaks but since determined to be absent or extremely rare include CBs: 20,23,30,34,35,36,38,50,53,72,73,78,79,80,90,107,108, 117,125,127,140,142,143,145,148,150,152,155,159,160,161,168,182.
Congener pairs are based on Fait et al. (1989). Where one of the pair is far removed from the sequence, the designation "w" (with) serves as a reminder. If other authors resolved the pair or reported alternate co-elutions, it should be obvious from the data entered.

[b] Fait et al. (1989) as revised by ED Pellizzari (personal communication, Dec 18, 1996). Workers were transformer repair workers currently exposed to PCBs while "general" included workers never directly exposed to PCBs. Concentrations based on median values or range if median was less than zero.

[c] Bush et al. (1985b). Forty samples from breast-feeding mothers at 3 weeks postpartum. Changes from published designations suggested by the senior author. Mean concentrations were converted to percent reported, accounting for 82% of total PCB.

[d] Wolff et al. (1982). Median concentrations in adipose tissue from 26 capacitor manufacture workers. Samples taken in 1976.

[e] Original designation changed to unidentified.

Appendix Table 6. Percent congener composition of airborne PCBs over a NPL landfill in Southern Illinois, over 2 areas near Green Bay, Lake Michigan, over Bloomington, Indiana and Chicago, Illinois.

B & Z No.[a]	Sangamo[b] Landfill	North[c] Green Bay	UnivWisc[c] Green Bay	Indiana[d]	Illinois[e]
4 + 10		0.64	1.36		1.42
5 + 8[f]	0.21				
6		1.66	1.23		1.59
7	0.05				
8 w 5[e]					
10 w 4					
12 + 13	0.05	1.07	0.37		
13 w 12					
15	0.91				
16+ 32	7.40	6.63	5.75	6.6	7.51
17 (w 18)[d]	2.18	2.85	2.74		
18	7.38	5.75	6.48	6.9	6.94
22	5.64			4.7	
24 + 27		1.09	1.67	0.4	1.31
25	0.29			0.3	0.74
26		1.58	2.20	1.1	1.73
27 w 24					
28 (+ 31)	19.89	14.07	13.49	19.8	15.63
29		0.27	0.60	0.2	
31 (w 28)					
32 w 16					
33 (+ 21 + 53)[e]	1.37	4.15	4.41	5.8	8.35

Appendix Table 6. Percent congener composition of airborne PCBs over a NPL landfill in Southern Illinois, over 2 areas near Green Bay, Lake Michigan, over Bloomington, Indiana and Chicago, Illinois.

B & Z No.[a]	Sangamo[b] Landfill	North[c] Green Bay	UnivWisc[c] Green Bay	Indiana[d]	Illinois[e]
37 (+ 42)	1.52	2.30	2.54		3.66
40	1.22	1.22	1.54	0.5	
41 (+ 71)[d,e]	4.48			0.2	3.87
42 (w 37)	2.90			1.9	
44	4.66	4.91	3.38	3.9	4.34
45	1.92	0.85	1.70	0.7	0.92
46		0.41	0.39	0.2	
47 + 48	2.87	3.15	2.37	2.0	1.90
48 w 47					
49	4.18	2.71	2.47	3.0	2.65
51		0.52	0.42	0.1	
52 (+ 43)[e]	6.42	5.96	5.20	4.8	3.17
56 + 60	2.22	2.52	2.36	2.9	2.50
60 w 56					
63		0.80	2.27	0.5	
64 (+ 71)[b]	3.53			1.9	
66 (+ 95)[c,e]	2.85	4.98	3.88	3.1	10.89
70 (+ 76)	3.35	4.68	3.93	4.2	3.65
71 (w 64 or 41)					
74	1.83	1.78	1.52	1.3	1.41
76 (w 70)					
77 (w 110)	0.23				

Appendix Table 6. Percent congener composition of airborne PCBs over a NPL landfill in Southern Illinois, over 2 areas near Green Bay, Lake Michigan, over Bloomington, Indiana and Chicago, Illinois.

B & Z No.[a]	Sangamo[b] Landfill	North[c] Green Bay	UnivWisc[c] Green Bay	Indiana[d]	Illinois[e]
81 (w 87)	0.01				
82	0.22	0.35	0.25		
83		0.10	0.12	0.1	
84 + 92	0.98	1.99	1.02	1.0	
85		0.47	0.31	1.2	
87 (+ 81)[c,e]	1.04	1.38	1.16	1.0	
91		0.70	0.42	0.6	
92 (w 84)	0.50				
95 (w 66)[c,e]	1.85			2.2	
97	0.65	0.86	0.97	0.8	0.69
99	0.87	1.34	0.84	0.9	
100		0.40	0.26	0.5	
101	2.59	2.84	3.02	3.9	1.67
105 (w 132,153)	0.18				
107 IUPAC 109[a]		0.03	0.13	0.01	
110 (+ 77)	1.89	1.74	1.75	1.8	3.06
114 (w 134)[c]	0.01				
118	0.59	0.83	0.69		1.21
119		0.07	0.08		
123					
126	0.02				
128	0.05	0.13	0.15	0.2	

Appendix Table 6. Percent congener composition of airborne PCBs over a NPL landfill in Southern Illinois, over 2 areas near Green Bay, Lake Michigan, over Bloomington, Indiana and Chicago, Illinois.

B & Z No.[a]	Sangamo[b] Landfill	North[c] Green Bay	UnivWisc[c] Green Bay	Indiana[d]	Illinois[e]
129 (w 178)	0.02				
131		0.10	0.12	0.01	
132 (w 105)	0.11				
134 (+ 114)	0.05	0.36	0.44	0.9	
135 + 124		0.36	0.51	0.7	
136	0.26	0.00	0.00	0.1	
137 (w 176)					
138 (+ 163)	0.41	1.34	1.36	1.0	1.48
141	0.07	0.28	0.27	0.3	0.23
146	0.08	0.23	0.43	0.2	
149	0.59	0.90	1.02	1.2	0.95
151	0.15	0.39	0.56	0.4	0.35
153 (+105,132)	0.41	2.36	2.64	1.8	2.40
156 + 171		0.23	0.30		0.15
157 + 200		0.00	0.14		
158		0.00	0.17	0.05	
163 (w 138)					
169		0.00	0.04		
170 + 190		0.25	0.83	0.1	0.25
171 w 156[c]				0.1	
172 + 197		0.03	0.06	0.01	
173		0.00	0.02		

Appendix Table 6. Percent congener composition of airborne PCBs over a NPL landfill in Southern Illinois, over 2 areas near Green Bay, Lake Michigan, over Bloomington, Indiana and Chicago, Illinois.

B & Z No.[a]	Sangamo[b] Landfill	North[c] Green Bay	UnivWisc[c] Green Bay	Indiana[d]	Illinois[e]
174		0.34	0.56	0.1	0.17
175		0.00	0.06	0.05	
176 + 137	< 0.01	0.03	0.26	0.1	
177		0.09	0.33	0.1	0.11
178 (+ 129)[c]	0.02	0.13	0.28	0.1	
180		0.65	0.74	0.2	1.30
182 (w 187)					
183		0.28	0.53	0.2	0.15
185		0.02	0.08	0.02	
187 (+ 182)		0.60	0.23	0.4	0.26
189		0.00	0.06		
190 w 170					
191		0.00	0.01		
193		0.00	0.08		
194		0.07	0.06		
195 + 208		0.60	0.31		0.13
196 + 203		0.27	0.95		
199 IUPAC[a]		0.02	0.03		
200 w 157[a]					
201 IUPAC[a]		0.33	0.41	0.1	0.16

Appendix Table 6. Percent congener composition of airborne PCBs over a NPL landfill in Southern Illinois, over 2 areas near Green Bay, Lake Michigan, over Bloomington, Indiana and Chicago, Illinois.

B & Z No.[a]	Sangamo[b] Landfill	North[c] Green Bay	UnivWisc[c] Green Bay	Indiana[d]	Illinois[e]
203 w 196[c]				0.1	0.19
208 w 195					
Total	NA[g]	328 pg/m^3	339 pg/m^3	NA[h]	>4ng/m^3

[a]Numbering according to Ballschmiter *et al.*, 1987. The revised Ballschmiter convention differs in that congeners 107, 108 and 109 are now referred to as numbers 109, 107 and 108, respectively. Congeners 199, 200 and 201 are now referred to as numbers 200, 201 and 199, respectively. Unresolved congeners are reported as X+Y for one and Y w (with) X for the co-eluting one(s). If one or more laboratories achieved resolution, co-elution is presented parenthetically; the Sangamo samples were analyzed by GC-MS so nonisomeric co-eluting congeners (e.g., 66/95 and 77/110) are resolved.

[b]High volume collection on XAD-2 resin after filtering dust (Hansen *et al.,* 1997). Only 52 of 96 congeners were above detection limits for GC-MS.

[c]Over water (NGB) and over land (UW-GB (Hornbuckle *et al.*, 1993).

[d]Bloomington, IN. Estimated from detailed figure (Panshin and Hites, 1994).

[e]Chicago, IL. Percentages derived from 40 most abundant peaks (Simcik *et al.*, 1998).

[f]Several authors have reported CBs 5+8 to account for as much as 30% of the total PCB in some areas (see Monosmith and Hermanson, 1996).

[g]Not available. Pooled extracts from 18 24 h high volume collections for bioassay.

[h]Not available. Averaged from combined data sets.

Appendix Table 7. Percent congener composition of PCBs in various plants.

B & Z No.[a]	Peat Moss[b]	Tomatoes[c] (market)	Tomatoes[c] N. Bedford	Lichens[d]	Vascular[e] Plants
5 + 8[d]				2.1	
6		1.9	0.5		
7					
8[c]		11.2	3.6		
15					0.1
16	0.64	4.3	1.2		
17 (w 18)[b]					
18 (+ 17)	9.25	14.1	3.3	1.0	0.1
22	1.00	1.8	1.8		
24 + 27	0.64				
25 (w 26)[b]					
26 (+ 25)	0.92				
27 w 24					
28 (+ 31)[b,d]	12.8	6.8	7.0	3.0	0.1
31 (w 28)		6.7	6.0		0.1
32 w 16[b]					
33	5.98	4.8	2.1	0.5	
37		5.9	1.7		
40	0.85				0.1
41 + 64[b]	2.06				
44	3.20	2.7	3.1		3.0
47 + 48	1.21				4.0
48 w 47					

Appendix Table 7. Percent congener composition of PCBs in various plants.

B & Z No.[a]	Peat Moss[b]	Tomatoes[c] (market)	Tomatoes[c] N. Bedford	Lichens[d]	Vascular[e] Plants
49	3.06	1.2	3.9		
52	8.42	6.9	5.6		2.0
56 + 60					2.5
60 w 56					
66 (+ 95)[d,e]	1.28	2.2	2.6	4.0	9.0
70 (+ 76)[d]	3.20	2.0	4.0	3.5	
74		0.9	1.8		0.5
77 (w 110)[c,e]	0.23				
84	0.4	1.0	1.02		
87 (+ 81)	0.92	1.1	1.6	+	6.0
91		0.70	0.42		
95 (w 66)	3.77	2.9	4.4		
97		0.4	1.7		3.0
99		0.5	2.9	+	
101 (+ 90)[b]	9.96	3.6	9.3	2.0	11.0
105		0	1.2		7.0
110 (+ 77)[c,e]	4.06	2.4	8.2	16.0	13.0
114 (w 134)					0.1
118	4.27	1.5	4.0	2.0	8.5
128	0	0.2	0.7		1.5
129 (w 178)					0.5
134 (w 114)					
135 (+ 144)[e]		0.7	1.1		1.5
136	0.71	0.4	0.5		

Appendix Table 7. Percent congener composition of PCBs in various plants.

B & Z No.[a]	Peat Moss[b]	Tomatoes[c] (market)	Tomatoes[c] N. Bedford	Lichens[d]	Vascular[e] Plants
137					0.5
138	4.84	2.8	5.0	11.0	8.0
141	1.14	1.0	1.2		1.0
146	0.28			1.0	
149	2.99	2.0	4.1	8.0	5.0
151	3.13	0.6	1.0	1.0	1.0
153 (132)	4.56	2.8	5.1	10.0	5.0
156 + 171		0.1	0.1		0.1
170 + 190	0	1.7	1.6	0.5	0.5
171	0.42				
173					0.1
174	1.14	0.4	0.9	0.5	0.3
176	0.07				
177	0	0.1	0.4	0.3	
178	0			2.0	
179	0.21				
180	1.14			5.0	0.6
182 (w 187)					
183	1.56			0.3	0.2
185					0.1
187	0.14			2.5	0.9
190 w 170					
191					0.1
194					0.1

Appendix Table 7. Percent congener composition of PCBs in various plants.

B & Z No.[a]	Peat Moss[b]	Tomatoes[c] (market)	Tomatoes[c] N. Bedford	Lichens[d]	Vascular[e] Plants
195 + 208					0.1
196 + 203					0.1
199 IUPAC[a]					0.1
206		0.2	0.4		
Total	140 ng/g	4.4 ng/g	2.2 ng/g	ca 8 ng/g	ca 5 ng/g

[a]Numbering according to Ballschmiter *et al.* (1987) as revised (Frame *et al.*, 1996; Guitart *et al.* 1993). Unresolved congeners are reported as X+Y for one and Y w (with) X for the co-eluting one(s) as in Appendix Table 6.

[b]Southern Finland (Himberg and Pakarinen, 1994).

[c]Non-local market vs. tomatoes ground near New Bedford Superfund site (Cullen *et al.*, 1996).

[d]Estimated from detailed figure (Muir *et al.*, 1993). Great Lakes region in SE Ontario.

[e]Estimated from detailed figure (Dushenko *et al.*, 1996). Canadian Arctic near military radar site.

Appendix Table 8. PCB congener composition of 1-hour high-volume air collections at 0.15, 0.6, and 0.9 m above the surface of an aged electrical dump.

B&Z No.[a]	Extract (ng/µl)			% 0.15 m	
	0.15 m	0.6 m	0.9 m	0.6 m	0.9 m
1	0.028	0.014	0.009	50.00	32.14
7	0.190	0.075	0.048	39.47	25.26
12+13	0.140	0.036	0.023	25.71	16.43
16+32	25.000	4.070	2.450	16.28	9.80
18	20.300	7.500	4.960	36.95	24.43
22	14.750	2.680	1.610	18.17	10.92
28+31	50.000	10.000	6.000	20.00	12.00
37	3.950	0.925	0.531	23.42	13.44
40	2.700	0.678	0.364	25.11	13.48
41+64	15.500	2.090	1.090	13.48	7.03
42	6.400	1.570	0.838	24.53	13.09
44	12.800	2.670	1.390	20.86	10.86
45	4.870	0.931	0.590	19.12	12.11
49	9.300	2.330	1.190	25.05	12.80
52	18.400	3.660	2.520	19.89	13.70
56+60	7.500	1.560	0.914	20.80	12.19
66[b]	5.400	1.570	0.866	29.07	16.04
70+76	7.850	2.090	1.120	26.62	14.27
77[b]	0.514	0.151	0.089	29.38	17.32
81[b]	0.025	0.008	0.005	32.00	20.00
82	0.523	0.175	0.099	33.46	18.93
84	1.760	0.572	0.331	32.50	18.81
87	2.640	0.844	0.502	31.97	19.02
91	1.300	0.395	0.222	30.38	17.08
92	1.140	0.375	0.221	32.89	19.39

Appendix Table 8. PCB congener composition of 1-hour high-volume air collections at 0.15, 0.6, and 0.9 m above the surface of an aged electrical dump.

	Extract (ng/µl)			% 0.15 m	
B&Z No.[a]	0.15 m	0.6 m	0.9 m	0.6 m	0.9 m
95	3.640	1.170	0.645	32.14	17.72
97	1.275	0.435	0.240	34.12	18.82
99	2.180	0.696	0.408	31.93	18.72
101	7.705	2.700	1.700	35.04	22.06
105	0.564	0.169	0.140	29.96	24.82
110	4.550	1.480	0.914	32.53	20.09
118	1.500	0.502	0.323	33.47	21.53
128+167	0.111	0.036	0.029	32.43	26.13
130[c]	0.063	0.014	0.015	22.22	23.81
134	0.111	0.047	0.026	42.34	23.42
132+168	0.247	0.082	0.051	33.20	20.65
136	0.615	0.191	0.113	31.08	18.39
138+163	0.903	0.315	0.216	34.88	23.92
141	0.174	0.055	0.040	31.61	22.99
146[c]	0.076	0.069	0.042	90.79	55.26
149	1.440	0.487	0.301	33.82	20.90
151[c]	3.800	0.124	0.073	3.26	1.92
153	0.824	0.317	0.198	38.47	24.03
158	0.099	0.041	0.023	41.41	23.23

[a]Concentrations of the following congeners or congener pairs were below the limit of quantitation in one or more samples: 5/8; 14; 15; 17; 33; 47/48; 74; 114; 119; 122; 123; all heptaCBs.

[b]These *ortho*-poor congeners differed markedly from the remaining homologs.

[c]Not included in averages for Table 4.3 due to unreliable quantitation.

Appendix Table 9. Proportions of PCB congeners in contaminated sediments of Woods Pond, MA suggesting dechlorination compared to sediment, water and biota congeners in Western Lake Erie.

	Woods Pond[a]		Western Lake Erie[b]					
B & Z No.	8WP Sed	WP-4 Sed	Sediment	Water	Plankton	Zebra Mussels	Gammarus	Cray-fish
16+ 32	0.1	0.3						
17	0.1	0.1						
18 + 15	0.2	0.2						
22 + 51	0.1	0.5						
25	0.1	0.2						
26	0.2	0.3						
28	0.3	0.3	1.8	3.6	1.3	0.7	0.8	0.8
31	0.3	0.2						
33; 20; 53	0.2	0.9						
37								
42 + 59	0.3	0.8	1.4	2.3	0.9	0.6	0.5	0.0
44 [104]	0.5	0.8	2.1	5.9	1.8	1.3	1.4	0.3
46		0.4						
47 + 48	0.4	3.4						
49	0.8	3.0	1.9	5.9	1.4	1.1	1.2	0.5
52	1.1	2.7	3.7	12.7	3.0	2.3	2.5	1.4
56 + 60	0.2	0.1	2.6	3.2	1.7	1.0	1.3	1.2
64; 41; 71	0.3	0.7	0.4	0.5	0.3	0.2	0.2	0.0
66 (+ 95)			6.1	11.8	6.4	3.5	3.5	2.3
70	0.7	0.6	3.1	7.7	1.9	1.3	1.2	0.3
74 + [94]	0.1	0.2	1.7	2.3	1.0	0.9	1.0	0.8
82	0.1	0.1						
83 + 112	0.2	0.3						
84 + 92	1.3	1.5						
85	0.4	0.4						
87;117;115	0.6	0.4	2.0	4.1	2.1	1.6	1.3	1.2

Appendix Table 9. Proportions of PCB congeners in contaminated sediments of Woods Pond, MA suggesting dechlorination compared to sediment, water and biota congeners in Western Lake Erie.

	Woods Pond[a]		Western Lake Erie[b]					
B & Z No.	8WP Sed	WP-4 Sed	Sediment	Water	Plankton	Zebra Mussels	Gammarus	Cray-fish
91	0.3	1.2						
95;66;[102]	2.4	2.6						
96	0.1	0.6						
97 + [152]	0.4	0.3	1.5	1.8	1.4	1.0	0.9	0.8
99	1.2	1.8	2.3	3.6	2.7	2.5	2.8	3.7
[100]	0.2	0.6						
101 + 90	3.9	3.6	4.6	8.2	6.1	4.5	3.7	4.1
103	0.2	0.6						
105 [w 132]			2.5	1.4	2.4	2.6	2.0	1.3
109 + 147	0.6	0.8						
110 + [148]	2.9	2.2	5.7	5.0	5.3	3.5	4.0	1.2
118 + 149	7.9	5.6	4.2	3.2	2.7	3.4	3.8	4.8
119 [+150]	0.3	0.7						
128	0.6	0.4						
129	0.3	0.2	0.8	0.0	0.9	0.9	1.1	1.2
[130]	0.7	0.5						
131	0.4	0.3						
132 + 105	1.7	1.0						
134 + 114	0.3	0.3						
135 + 124	2.0	1.5						
136	1.2	1.2						
137	0.1	0.1						
138;163;164	8.8	6.6	9.0	3.6	9.3	11.2	11.2	13.9
140		0.5						
141	1.6	1.1	1.5	0.5	1.8	1.8	1.4	1.9
144	0.6	0.5						

Appendix Table 9. Proportions of PCB congeners in contaminated sediments of Woods Pond, MA suggesting dechlorination compared to sediment, water and biota congeners in Western Lake Erie.

	Woods Pond[a]		Western Lake Erie[b]					
B & Z No.	8WP Sed	WP-4 Sed	Sediment	Water	Plankton	Zebra Mussels	Gammarus	Cray-fish
146	2.0	1.7	1.1	0.5	1.7	2.0	2.1	2.6
149 w 118			5.4	3.6	10.4	7.4	6.7	4.4
151	2.3	2.4	1.6	1.4	3.6	2.5	2.7	2.1
153	10.5	8.3	5.4	2.7	8.4	11.1	10.5	13.5
154	0.3	0.6						
156 + 171	1.3	1.0						
158	0.7	0.5						
167	0.8	0.5						
170	2.8	2.2	3.5	0.5	2.2	3.6	3.7	3.8
171 w 156			0.7	0.0	0.8	1.2	1.2	1.1
172	1.0	0.7	0.4	0.0	0.3	0.5	0.6	0.6
173	0.1							
174	3.6	3.0	2.3	0.5	2.6	3.0	2.8	2.4
175	0.4	0.3						
176	0.3	0.3						
177	2.0	1.8						
178	0.7	0.7						
179	1.5	1.6						
180	7.9	6.9	5.6	1.4	4.8	8.1	8.4	10.8
182			3.2	0.9	4.3	6.1	6.2	8.1
183	2.1	1.9	1.4	0.5	1.8	2.6	2.5	2.3
185	0.4	0.4	0.2	0.0	0.3	0.2	0.2	0.2
187	4.1	3.9						
190	0.6	0.6						
193	0.4	0.5						
194	1.1	1.1	4.9	0.0	0.5	1.0	1.5	1.2

Appendix Table 9. Proportions of PCB congeners in contaminated sediments of Woods Pond, MA suggesting dechlorination compared to sediment, water and biota congeners in Western Lake Erie.

	Woods Pond[a]		Western Lake Erie[b]					
B & Z No.	8WP Sed	WP-4 Sed	Sediment	Water	Plankton	Zebra Mussels	Gammarus	Cray-fish
195	0.6	0.6	0.9	0.0	1.2	0.6	0.5	0.6
196 + 203	1.4	1.5						
197	0.1							
199	1.3	1.3						
200			0.5	0.0	0.4	0.5	0.5	0.5
201		0.2	2.1	0.9	1.4	2.2	2.0	2.5
203 w 196			1.3	0.0	0.8	1.4	1.3	1.0
205	0.1							
206	0.6	0.7	0.7	0.0	0.1	0.3	0.7	0.2
Total PCB (ppm)	123	155	0.109	$0.22\ E^{-6}$	0.028	0.064	0.081	0.046

[a]Representative sediment samples subjected to extensive specific congener analysis (Bedard and May, 1996). Relative concentrations in Mol %. Bracketed congeners are those not normally present in Aroclors (Aroclor 1260 was the main source) and presumed due to dechlorination of other congeners.

[b]Observed concentrations used to validate a nonequilibrium, steady-state biota-sediment accumulation model (Morrison et al., 1996). Absolute concentrations published were converted to weight % for consistency among other Appendix Tables.

Appendix Table 10. Percent composition (CBs reported) of PCB residues in fish from the Great Lakes.

B & Z No.[a]	Lake Ontario[b]		Lake Erie[c]		Lake Superior		
	Trout	Salmon	Silver Bass		White Fish[d]		Trout[e]
	Whole	Whole	Muscle	Liver	Muscle	Liver	Whole
8					6.79	20.78	11.43
15							12.57
16					0.41	0.74	
18							2.40
25					0.34	0.20	
28	0.72	1.12	0.13	0.13	25.00	1.09	
+31			0.01	0.07			
33					0.88		8.29
40							1.71
42			0.04	0.46			
44			0.58	0.62	0.68	1.49	
47(48)	1.40	1.42					
49			0.56	0.60			3.43
52	1.49	1.58	1.16	1.22	0.41		3.71
56+60	1.85	1.78	2.56	2.52			
63					7.81	8.28	
64			0.16	0.15			
66(95)	3.54	4.27	3.33	3.50	0.61	1.09	
70(76)	3.21	4.17	1.00	1.09	1.29	2.03	
74			0.54	4.00		1.24	
77(110)	0.18	0.41	0.00	0.00			
81(?)[f]	*0.90*	*1.32*			*1.97*	*1.93*	
82					3.46		
84	6.38	5.49			1.63	1.24	
87			1.69	1.88			
87(97)	3.91	6.86					
89						4.71	

Appendix Table 10. Percent composition (CBs reported) of PCB residues in fish from the Great Lakes.

B & Z No.[a]	Lake Ontario[b]		Lake Erie[c]		Lake Superior		
	Trout	Salmon	Silver Bass		White Fish[d]		Trout[e]
	Whole	Whole	Muscle	Liver	Muscle	Liver	Whole
95	1.71	2.13					3.71
97			3.16	1.56	0.20	0.89	
99			2.91	3.54	0.82	0.99	
101	6.11	6.50	6.74	5.67	0.88	0.89	6.29
105(132)	2.54	2.44	1.98	2.32	2.11	2.08	
110(77)	4.77	5.79	4.57	4.81	1.56	3.32	
114					3.87		
118	6.36	5.08	3.33	4.02	1.77	1.84	6.00
128	1.74	1.52			0.41	0.69	
129			1.29	1.46	1.43	1.69	
136					0.34	0.99	
137					2.79	5.46	4.29
138(163)	6.03	5.49	9.79	12.37	4.14	4.12	16.00
141	1.97	1.68	2.01	2.33	0.88	2.68	
146			2.20	2.55	1.15	1.19	
149	3.28	4.93	7.60	5.21	1.49	1.19	3.14
151					0.54	1.04	
152			1.98	3.66			
153	10.84	7.77	10.98	6.60	5.23	4.41	
156(171,202)					0.48	0.64	
167					1.29	0.94	
169			0.00				
170+190	2.07	1.63	3.51	1.44	1.70	1.64	
171			0.62	0.81			
172			0.61	0.76	0.27	0.60	
173					0.34	0.20	
174			2.67	3.07	0.68	0.84	

Appendix Table 10. Percent composition (CBs reported) of PCB residues in fish from the Great Lakes.

B & Z No.[a]	Lake Ontario[b]		Lake Erie[c]		Lake Superior		
	Trout	Salmon	Silver Bass		White Fish[d]		Trout[e]
	Whole	Whole	Muscle	Liver	Muscle	Liver	Whole
177					0.82	1.34	
179					0.95		
180	4.73	3.96	8.56	3.87	2.24	2.23	7.14
182+187							
183	1.65	1.42	2.54	3.13	1.43	1.54	
185			0.28	0.45			
187	2.86	2.64	5.46	6.66	2.58	3.37	5.14
191					3.94	1.84	
194	0.56	0.46	1.01	1.40	0.34	0.35	1.60
195			0.33	0.43		2.58	
196+203	1.25	1.02			1.02	1.88	
199(201)	1.10	0.91	2.24	2.91	1.02	1.39	3.14
200			0.43	0.81			
203			1.32	1.75			
206	0.24	0.05	0.12	0.17		0.30	0.00
209	0.16	0.00					

[a]See Tables 1 or 2 for structures.
[b]Niimi and Oliver (1989).
[c]Koslowski et al., 1994.
[d]Gerstenberger et al., 1997.
[e]Swackhammer and Hites, 1988.
[f]Probable interference or mis-assignment.

Appendix Table 11. Values used to calculate "average" fish residue profiles for Chapter 8 and Table 8.2.

B&Z No.	Mean %[a]	Striped Bass[b]	Tuna[c]	Eel[d]	Flounder[e]
8*[a]	0.49			0.3	
16	0.45			0.5	
17	0.20			0.2	
18	0.35	0.3		0.4	
22	0.50			0.5	
25	0.29			0.25	
28(31)*[a,f]	1.03	1.4	0.53	1.6	0.23
(31)	0.69	1.9		0.15	
33	0.54			0.2	
40	0.48		0.27	0.7	
42	1.61	2.8		2.0	
44	1.54	2.3	1.87	3.6	0.21
47(48)	2.80	4.0	0.40	9.2	0.34
49	2.48	4.7	0.40	6.4	0.33
52	1.94	4.4	0.67	4.9	0.92
56 + 60	1.75			1.9	0.69
63*[a]	0.47			0.15	
64(41)	1.57	4.1	0.80	2.3	0.48
66(95)	3.15	2.4	3.48	5.1	2.44
70 + 76	1.78	2.4	0.67	0.2	1.33
74	1.13	1.6	0.53	1.2	1.78
77	0.20				
81??[g]	*1.40*				
84	3.85	2.4	6.55		0.63
85	1.83	1.9		0.7	2.89
87	3.12		2.14	1.0	
87(97)	5.39				
91	1.49			2.3	0.69

Appendix Table 11. Values used to calculate "average" fish residue profiles for Chapter 8 and Table 8.2.

B&Z No.	Mean %[a]	Striped Bass[b]	Tuna[c]	Eel[d]	Flounder[e]
95	2.43				3.44
97	1.52	1.5	1.47	1.9	0.87
99	3.59	2.8	1.74	6.7	6.55
101	4.31	3.0	1.74	5.1	4.44
105(132)	2.94	0.8	5.75	0.8	7.10
110(77)	5.01	3.8	5.88	7.9	5.77
114*[a]	0.39				
118	5.40	2.5	8.16	5.8	10.21
128	1.97	1.1	4.14	0.9	4.00
129(178)	1.01			0.3	
136	0.86	2.4		0.25	0.43
137(176)*[a]	0.29			0.3	
138(163)	7.44	4.5	13.37	4.0	12.21
141	1.30		1.47	0.6	0.47
146	1.61		1.74	1.3	1.66
149	3.60	1.6	1.60	3.3	4.99
151	1.34	1.3	3.34	0.5	1.03
153	8.55	7.0	6.02	9.6	10.99
156(171)	1.34		2.41	0.5	2.00
170 + 190	2.17	1.2	5.48	0.45	1.33
171	0.61				0.60
172	0.32				0.08
173	0.34				
174	1.61		4.41	0.05	0.23
177	1.00		2.67	0.15	0.36
179	0.61				0.27
180	3.78	3.1	4.28	0.6	2.77
183	1.35	1.0	1.07	0.5	1.22
185	0.16				0.03

Appendix Table 11. Values used to calculate "average" fish residue profiles for Chapter 8 and Table 8.2.

B&Z No.	Mean %[a]	Striped Bass[b]	Tuna[c]	Eel[d]	Flounder[e]
187 + 182	2.59	1.7	3.34	0.7	1.44
194	0.45	0.6	0.53	0.05	0.05
195	0.29			0.5	0.03
196 + 203	0.86		0.40	0.15	
199	0.87		0.67	0.1	0.06
200	0.43				
206	0.11				0.010
209	0.05				0.002

[a] Includes Lake Ontario trout and salmon (Niimi and Oliver, 1989), Lake Erie silver bass muscle (Koslowski et al., 1994), and Lake Superior white fish muscle (Gerstenberger et al., 1997) from Appendix Table 10. Values for some congeners in white fish were adjusted and are indicated by *.

[b] Striped bass from Hudson Estuary (Bush et al., 1989). Major congeners (80 % of total PCB).

[c] Tuna muscle from the Mediterranean (Porte and Albaiges, 1993). Values converted from figure normalizing congener percents to CB 138.

[d] Eel from New Bedford Harbor (Lake et al., 1995). Values derived from figure of percent composition.

[e] Flounder from the Gulf of Gdansk (Falandysz et al., 1997). Percentages calculated from ng/g lipid. Values for 13 congeners (about 3% total PCB) were not included.

[f] Congeners are listed with co-eluting peaks; if one or more authors resolved the congeners or did not list co-eluting congeners the values are parenthetic.

[g] Although a peak was identified as CB 81 in two studies, this proportion of a very rare CB probably indicates erroneous assignments, perhaps interference from CBs 85 and/or 87.

Appendix Table 12. Crude collective "average" PCB congener composition of human breast milk. Taken from Hansen, 1998 a,b,c with some values (Wales) adjusted for more recent data.

B&Z	AVG %[a]	MI	CAN	WALES	NOR	N.Y.	FIN
1	+					0.75	
4	+					0.75	
8	+					0.75	
18	1.8			3.2			0.31
22	+	0.6					
28	3.6	8.8	1.87	6.5	2.10	2.42	2.03
31	+					0.75	
33	1.8	2.2	3.0				0.21
37	2.1	2.9	3.33				
41	1.1	1.3	0.83				
44	1.5	0.8	2.2	2.1		0.75	
46	+	0.25					
47	0.3	0.4					0.18
49	1.3	0.7	2.0				
51	+						0.27
52	1.7	1.9	1.6	3.3			0.18
60	1.4	0.7	2.91			1.77	0.33
66	1.5		1.66	2.9			0.86
70	+	0.6					
74	5.5	11	3.74	4.0	3.39	6.42	5.00
77	0.01				0.01		0.011
84	+					0.25	

Appendix Table 12. Crude collective "average" PCB congener composition of human breast milk. Taken from Hansen, 1998 a,b,c with some values (Wales) adjusted for more recent data.

B&Z	AVG %[a]	MI	CAN	WALES	NOR	N.Y.	FIN
87	0.7	0.8	0.62				
95	+						
99	4.5	4.8	5.82	4.1	3.63	4.15	4.48
101	1.0	1	1.04	1.8	0.30	1.02	0.34
105	2.4		1.9	2.7	2.07	3.66	1.80
109	+?	0.3					
110	1.1	1	1.5	1.1		1.74	0.20
114	0.6	0.3			1.08		0.44
118	7.1	6.5	4.16	4.2	7.04	9.43	11.32
119	1.0	0.1		1.9			
122	+	0.5					
123	+						0.38
126	0.04			0.04	0.04		0.037
128	0.9	0.3	1.04	0.8	0.99	2	0.26
129	+		0.62				
130	+	0.6					
132	+						
135	+	0.5					
137	1.7	0.9	2.9				
138	12.4	10	8.0	13.1	23.33	7.92	16.55
138*[b]	10.0	7.5	6.0	9.9	17.5	6.94	12.41
141	0.2	0.3	0.21		0.05		0.15

Appendix Table 12. Crude collective "average" PCB congener composition of human breast milk. Taken from Hansen, 1998 a,b,c with some values (Wales) adjusted for more recent data.

B&Z	AVG %[a]	MI	CAN	WALES	NOR	N.Y.	FIN
146	2.0	1.9				2.11	
149	1.9			1.9			
151	0.7	0.6	0.7	0.9			
153	20.1	12	25.8	15.8	30.75	12.1	24.35
156	3.1	4.9	1.46	3.0	3.12	3.77	2.63
157	0.4	0.5	0.42		0.43		0.35
158	0.7	0.6				0.83	
163	+					0.98	
163*[b]	3.1	2.5	2.0	3.3	5.8	0.98	4.10
167	0.7	0.8					0.66
169	0.03				0.05		0.017
170	4.4	5.3	2.29	4.5	6.64	2.34	6.41
171	1.3	0.4				2.15	
172	+	0.8					
174	+	0.4					
177	+	0.6					
180	9.4	5.3	5.8	14.6	13.60	4.53	12.27
183	1.5	1.4	2.0	1.3		0.83	2.04
185	+	0.1					
187	3.2	1.5	2.4	4.2			4.61
189	1.1	2.4	0.21			1.55	0.18
191	0.7	0.9	0.42				

Appendix Table 12. Crude collective "average" PCB congener composition of human breast milk. Taken from Hansen, 1998 a,b,c with some values (Wales) adjusted for more recent data.

B&Z	AVG %[a]	MI	CAN	WALES	NOR	N.Y.	FIN
193	0.6	0.2	1.04				
194	1.0	0.5	1.2	0.9	1.29		1.15
195	0.3	0.3					
196	0.2	0.2					
199	1.1	0.8	1.25	1.4		0.72	
203	0.9	0.8	0.83			1.06	
205	0.1	0.1					
206	0.1	0.2	0.21		0.11		0.03
209	0.4	0.1	0.62				

[a] Since numbers of values are highly variable, the average sum will be different than 100%. Estimated low values (NY) are considered "present" (+) if not reported in other studies. The single sample from Michigan is likewise "present" (+) if more extensive surveys did not report a value.

[b] Values for CB 138/CB 163 were adjusted according to literature proportions. Similar operations could be performed for CBs 66/95 and CBs 105/132.

Appendix Table 13. Frequencies of congener occurrence in breast milk surveys from Canada (Mes et al., 1993; Newsome et al., 1995) and from Finland (Vartiainen et al., 1997; Kiviranta et al., 1998) and degree maximum levels exceed means in survey from Wales.

B & Z No.	Canada (% detected)		Finland (% detected)				Wales (max/ mean)[d]
			Urban		Rural		
Year	1986	1992	1992	1987	1992	1987	1990
(n)	(412)	(497)	(14)	(47)	(28)	(37)	(115)
18			14	45	89	22	9.9
28 + 31	79	85	100	100	100	98	6.3
33 (+ 53)	25	14	7	49	89	30	
37 + 42	16	16					
40		14					
41 (+ 71)		17					
44	NR[a]	51					8.2
47 (+ 48)			100	87	100	89	
49	NR[a]	98	7	0	32	0	
51			7	11	75	0	
52	39	22	57	89	93	89	9.1
60 (+ 56)	54	54	100	91	100	100	
66 (+ 95)	83	73	100	100	100	100	13.8
74	90	98	100	100	100	100	3.3*
77 (w 110)			93	21	54	27	
81				0		0	
87	18	<100[b]					
99	93	90	100	100	100	100	8.8*
101 (+ 90)	64	47	100	94	93	43	8.8
105 + 132	19	91	100	100	100	100	10.4
110 (+77)	14	20	93	94	54	30	10.9

Appendix Table 13. Frequencies of congener occurrence in breast milk surveys from Canada (Mes et al., 1993; Newsome et al., 1995) and from Finland (Vartiainen et al., 1997; Kiviranta et al., 1998) and degree maximum levels exceed means in survey from Wales.

B & Z No.	Canada (% detected)		Finland (% detected)				Wales (max/ mean)[d]
			Urban		Rural		
Year	1986	1992	1992	1987	1992	1987	1990
114 (+ 134)			100	100	100	100	
118	83	99	100	100	100	100	9.0
122			7	0	0	0	
123			100	100	100	100	
126			100	91	100	100	
128 + 167	80	44	100	94	100	89	5.2
129 (+178)	68	24					
136	0.7						
137	23	99					
138(+163)	95[c]	100	100	100	100	100	2.7*
141	49	26	100	8.5	18	5	
149							28.3
151	74	18					13.4
153	95	100	100	100	100	100	3.5*
156 (+202)	93	98	100	100	100	100	7.1*
157	72	52	100	98	100	100	
159				0		0	
167 w 128?			100	100	100	100	
169			100	85	100	95	
170 (+ 190)	85	99	100	100	100	100	3.5
180	95[c]	98	100	100	100	100	2.8*
183	91	89	100	100	100	100	3.2*

Appendix Table 13. Frequencies of congener occurrence in breast milk surveys from Canada (Mes et al., 1993; Newsome et al., 1995) and from Finland (Vartiainen et al., 1997; Kiviranta et al., 1998) and degree maximum levels exceed means in survey from Wales.

B & Z No.	Canada (% detected)		Finland (% detected)				Wales (max/ mean)[d]
			Urban		Rural		
Year	1986	1992	1992	1987	1992	1987	1990
185		23					
187 (+ 182)	93	99	100	100	100	100	10.3
189	52	23	100	94	100	100	
191	40	26					
193	28	50					
194	90	86	100	100	100	100	4.7
201 (199)	87	90					8.8*
203 + (196)	90	92					
206	62	30	100	79	100	89	
209	25	14	100	0	54	0	
Total PCB (ng/g fat)	241	238	296	496	198	396	522

[a] Omitted due to unreliable results.
[b] Present in original report, but lost from data set.
[c] Poor recoveries (64%); not corrected for recovery.
[d] Ratio of reported maximum/reported mean (Duarte-Davidson et al., 1994). A non-detect in the reported range indicated less than 100% frequency; ratio* = congeners which were detected in all samples (i.e., the minimum was > 0).

Appendix Table 14. Abbreviations

ADD	Average Daily Dose
Ah	Aryl hydrocarbon
AhR	Aryl hydrocarbon receptor
B&Z No.	Ballschmiter and Zell congener numbering system
BAF	Bioaccumulation factor
BCF	Bioconcentration factor
BMF	Biomagnification factor
BSAF	Biota-Sediment accumulation factor
BROD	Benzoxyresorufin *O*-dearylase
CA	Chlorinated aromatic
CAS	Chemical Abstracts Service (CAS Numbers)
CB	Chlorobiphenyl (a specific PolyCB congener)
CBz	Chlorobenzene
CYP	Cytochrome P-450 monooxygenase; P_{450}
DDE	1,1-Dichloro-2,2-*bis*(4-chlorophenyl)ethylene; the 4-chloro is ofter designated *para*. The 4,4'- and 2,4'- (*p,p'*- and *o,p'*) are the most common isomers.
DDT	1,1,1-Trichloro-2,2-*bis*(4-chlorophenyl)ethane; "Dichloro-diphenyl trichloroethane". As with DDE, 4,4'- and 2,4'- (*p,p'*- and *o,p'*) are the most common isomers.
EROD	Ethoxyresorufin *O*-deakylase
GC	Gas chromatography
GSH	Glutathione
GST	Glutathione-S-Transferase
HCH	Hexachlorocyclohexane; lindane; "Benzene Hexachloride" or "BHC"
HRGC	High resolution gas chromatography
IUPAC	International Union of Pure and Applied Chemistry
K_{ow}	Octanol:Water partition coefficient
k_a	First-order absorption rate constant
k_e, k_{el}	First-order elimination rate constant
LADD	Lifetime Accumulated Daily Dose
MeSO	Methyl sulfonyl (methyl sulfoxide or sulfone metabolites)
NHATS	National Human Adipose Tissue Survey
PAH	Polycyclic aromatic (hydrocarbon); PNA has more uses
PBPK	Physiologically-based pharmacokinetics
PoBPK	Population-Based Pharmacokinetics
PCB	Polychlorinated biphenyl (generic term; mixture)
PCDD	Polychlorinated dibenzo-*p*-dioxin
PCDF	Polychlorinated dibenzofuran
PCN	Polychlorinated naphthalene
POP	Persistent organic pollutant
PROD	Pentoxyresorufin *O*-deakylase
RRT	Relative retention time

Appendix Table 14. Abbreviations

RyR	Ryanodine (-sensitive calcium channel) receptor
SAR	Structure:Activity relationship
ST	Sulfotransferase
$t_{1/2}$	Biological half-life
TCDD	2,3,7,8-tetrachlorodibenzo-*p*-dioxin; "dioxin"
TEF	TCDD equivalency factor; "T" often defined as "toxic"
TCDD EQ	TCDD equivalents (more accurate than "toxic" equivalents)
TEQ	TCDD equivalents; often defined as "toxic" equivalents
UDPGT	Uridine-disphospho-glucuronosyl transferase

INDEX

abridged 9, 21, 27, 32, 33, 144, 151, 154, 160
absorption 50, 101, 102, 107, 108, 111-114, 122, 123, 129-131
air 12, 32, 35-38, 40-49, 60-62, 65-70, 101, 133, 137, 138, 141
152, 153, 231, 241
alewife 57
algae 54, 56, 78, 98
anaerobic dechlorination 32, 40, 53, 64, 68, 69
analyte selection 32-33,126, 149, 152
Antarctic 51, 67
antiestrogenic 19, 82, 144
Arctic 51, 98, 136, 137, 240
arene oxide 75, 82-85, 87, 95, 96
Aroclor 1-4, 6, 7, 9, 14, 15, 22-24, 26-28, 32, 40, 44-47, 52, 54, 68, 72,
78, 81, 98, 99, 109, 111-115, 120-123, 143, 246
arthropods 75, 128, 130
Atlantic 37, 52, 61, 62, 67
atropisomer 13, 21
average daily dose 156, 260
Ballschmiter and Zell 4, 29, 31, 260
bass 71, 247, 252
bioaccumulation, BAF 55, 56, 70-72, 101, 128, 131, 133
bioconcentration, BCF 7, 21, 70, 71, 78, 122
biological action 16-19, 41, 142-145, 148
birds 72, 78, 80, 125-128, 133
blood 18, 27, 84-86, 93, 101, 103-107, 110-115, 118-120, 144
breast milk 13, 24, 31, 81, 114, 121, 141, 142, 146-149, 151
153-160, 253, 257
calcium 16, 19, 143, 261
capillary column 24, 28, 33
carrot 48, 141
catechol 87-89, 98, 144
catfish 27, 122, 123
chicken 27, 56
children (see also infant, developmental) 117, 143, 148
chiral 13
chitin 66, 68
Clophen 1, 72, 116
compartment 35, 52, 53, 55, 69, 101-103, 105, 107-111, 114, 120
130, 131, 157, 158
congener pairs 10, 16, 121, 230, 242
28/31 . . . . . 41, 43, 127, 134, 241, 250, 257
66/95 . . . . . 10, 16, 40, 63, 73, 127, 149, 150, 151, 236, 247, 250, 256
77/110 . . . . . 10, 40, 151, 236, 247
105/132 . . . . . 10, 16, 121, 136-137, 151, 153, 248, 251, 256, 257

138/163 . . . . 43, 61, 63, 73, 78, 149, 150, 152, 153, 242, 248, 251
182+187 . . . . 127, 249
conjugate 93, 94
"constants" 7, 12, 101
copepod 73
crayfish 75
cricket 179, 186
crustaceans 75
cuticle 49, 130, 131
cuticular penetration 50
CYP 17, 72, 74, 78, 79, 84, 86, 87, 90-93, 120, 125, 158
CYP 1A 72, 74, 86, 87, 90-92, 158
CYP 1B1 . . . . 90, 92
CYP 2A 17, 92
CYP 2B 72, 74, 78, 92
CYP 3A 72, 90-92
CYP 4A 90-92
DDE 24, 26, 27, 30, 94, 113, 115, 117, 144-146
DDT 37, 60, 146
dechlorination 14, 32, 40, 53, 64, 66, 68, 69, 80, 86, 116, 243, 246
detection 4, 16, 22, 33, 84, 98, 115, 121, 143, 148, 149, 236
developmental 16-19, 120, 142, 143, 156
diet 78, 79, 116, 121, 125, 126, 129, 133, 134, 137, 143, 151, 152, 156
diol 87-89
disposition 6, 35, 79, 82, 83, 90, 101, 107, 111, 112, 116, 118, 120, 122, 129-131, 159
dog 86, 92, 106, 118
eel 124, 125, 252
eggs 23, 72, 81, 126-128, 141
elimination . . 56, 71, 84, 94, 101, 102, 104, 105, 107-111, 114-118, 122-125, 129, 130, 145
elution 9, 236
enantiomers 13
estrogenic 19, 144
excretion 50, 81, 101, 107, 108, 110, 128, 130, 145
feces 93, 95, 108
fetus 82, 87, 94
First-order 54, 102-105, 111, 260
fish 28, 42, 51, 53, 56, 58, 71, 72, 78-80, 91, 92, 97, 98, 112-114, 122-126, 129, 133, 134, 137, 138, 141, 148-154, 243, 247, 250, 252
fish meal 113, 114, 134-141, 145
flies 49, 129-131
flounder 73, 137, 252
frequency 75, 126, 145, 148-150, 259
fugacity 12, 35, 36, 51, 54, 57, 58, 69, 71

gas chromatography, GC 4, 7, 9-11, 22-24, 28, 30, 40, 55, 110, 112, 117, 236
Gibbs free energies 14, 211, 213, 220
global fractionation 37, 41-44, 47, 192
grass 47-48, 75
great tit 128
growth dilution 55, 56, 121, 123, 148
half-life 101-131
Henry's Law 11, 12, 61, 67, 211, 220
herbage 41, 47, 49, 133, 137, 152
herring 73, 126, 137
human . . . 1, 13, 24, 27, 29, 33, 51, 57, 78, 82, 84-86, 92, 113-115, 118, 134, 137, 141, 142-154, 156, 157, 159, 227, 253, 260
hydroxy 38, 58, 82, 84, 90, 94, 98, 99, 144, 159
induction 17, 78, 84, 90-97, 112, 114, 120, 125, 131, 144
infant 121, 154, 156, 157
inhibition 90, 99, 131
insects 49, 50, 58, 92, 125-129, 133
invertebrates 50, 68, 81, 126, 128, 133
IUPAC 4, 7, 8, 29, 31, 96, 233, 260
Kanechlor 1
Lake Baikal 72
Lake Erie 67, 71, 149, 243, 252
Lake Huron 149
Lake Michigan 64, 68, 149, 231
Lake Ontario 23, 68, 252
Lake Superior 247, 252
landfill 40, 44-47, 49, 125, 231
lettuce 49, 141
lichens 48, 239
lobster 148
Log P 12
lung 29, 90, 95, 105, 107, 121
lymph, lymphatic 111, 112, 194
marine 42, 52-54, 58-60, 68-73, 76, 78-80, 125, 128, 133, 134, 137
marine mammals 80, 125, 134-137
mass balance 68, 82, 83, 97, 98, 120
mass spectroscopy 22
medical 33, 121, 142-144, 146, 148, 149
metabolic slopes 71, 84
metabolism . . . 6, 17, 50, 55, 57, 68, 71-75, 78-87, 90, 91, 93, 94, 97-99, 101, 105, 107, 111, 112, 118, 120, 128, 131, 133, 137, 145, 148, 159
metabolites 16, 21, 50, 58, 68, 80-84, 86-90, 93-99, 106, 108, 122, 129, 130, 133, 142, 143, 144, 154, 158-160, 260
methyl sulfonyls 88, 89, 95, 96, 98, 144, 158-160
milk (see breast milk) 257
mink 85, 88, 95

monkey . . . 106, 118
mosquito . . . 98, 129
mosquito fish . . . 98, 129
moss . . . 48, 49, 239
mullet . . . 78, 79
mussel . . . 66, 73, 75, 78, 137, 243
New Bedford . . . 28, 65, 66, 114, 141, 150, 240, 252
non-Hodgkins lymphoma, NHL . . . 145, 151
NIH shift . . . 75, 84-88, 95, 96
ocean . . . 35, 37, 51-53, 61, 62, 67
omnivorous . . . 134, 137, 151
packed column . . . 22-28, 30, 44, 110, 112
PAH . . . 60, 155
partition coefficients . . . 7, 10, 12, 21, 41, 70-72, 118, 119, 160
pattern . . . 13, 15, 22, 23, 29, 69, 116, 134, 149
PBPK . . . 103, 105, 118-120, 260
PCDD . . . 42, 44, 46, 48, 50, 155
PCDF . . . 48, 155
PCN . . . 155, 260
pharmacokinetics . . . 101
pigeon . . . 86
plankton . . . 67, 78, 243
plasma . . . 84, 85, 101, 107, 110, 111, 144, 159
porphyria . . . 18
porpoise . . . 73, 78, 137
potato . . . 49, 141
precipitation . . . 41
produce . . . 6, 23, 41, 49, 137-141, 152, 153
production . . . 1-3, 14-16, 35, 52, 57, 86, 90, 97, 98, 137, 138, 144, 221
profiles . . . 11, 14, 15, 17, 21, 23, 24, 27-29, 31, 32, 35, 41, 44, 45, 49, 53, 61, 62, 65, 68, 69-72, 75, 78-80, 83, 90, 97, 114, 121, 122, 125, 126, 128, 129, 131, 133, 134, 137, 141-149, 151, 152, 154, 155, 157-161, 250
pulse, pulsatile . . . 69, 80, 117, 137, 141, 149, 153-155, 157-159, 160
quail . . . 78
quinone . . . 144
resolution . . . 9, 21-24, 26-33, 59, 73, 104, 107, 114, 117, 144, 236
retention time . . . 4, 7-10, 22, 24, 26, 27, 30, 113, 115, 117
ryanodine, RyR . . . 16, 143
salivary gland . . . 93
salmon . . . 57, 247, 252
sculpin . . . 57, 72
sea . . . 74, 78, 136, 137
seal . . . 72, 78, 84, 85, 134-137, 152
sediment . . . 36, 38, 40, 64-66, 68, 69, 71, 133, 243, 246, 260

serum 78, 82, 114, 116, 145, 146, 148-151, 159
sheep 27, 44, 98, 110-115, 141
shrimp 73, 75
sink 60, 69
smelt 57
snail 98
soil 36, 40, 42, 45-47, 49, 69, 70, 129
solubility 4, 12, 39, 64, 93, 159
sporadic (exposure) 16, 58, 72, 120, 121, 128, 137, 141, 149 151-153, 158, 159
starling 125
steady state 54, 55, 65, 67, 69, 107, 109, 111, 112, 125, 134, 141, 142, 144-146, 148, 151
steroid 17, 18, 90-95
swallow 71, 125-128
swine 27, 56, 109, 111-115, 120, 121, 141, 143, 156
TCDD 6, 11, 16-19, 31, 46-49, 92, 137, 142-145, 151, 152, 154, 160
TEQ 15, 17, 44, 46, 47, 84, 87, 137, 142-145, 151, 155, 261
thyroid 17, 19, 90, 95, 143, 144
tomato 141
toxicant 101, 103, 107, 133, 143
toxicity 16-19, 33, 69, 86, 121, 154, 155, 159
toxicokinetics . . . . . viii, 101, 103, 112, 118, 120-122, 129, 131, 133, 154, 157
toxicological 33, 141, 154, 155, 157
trout 114, 122-125, 247, 252
tuna 78, 79, 252
urine 83, 93, 94, 108
vapor phase 37, 38, 40-45, 47-50, 63, 70
vapor pressure 11, 12, 39
vegetables 49, 137-141, 152
vitamin 16, 18, 143
volatile 11, 38, 40, 44, 47-49, 53, 61, 64, 66, 69, 153
volatility 11, 12, 22, 35, 42, 45, 47, 64
volatilization 3, 40, 45, 66, 69, 130
walrus 137
Webb and McCall 23, 24, 26-28, 113
whale 98
Yu-Cheng, Yusho 84, 118

GPSR Compliance
The European Union's (EU) General Product Safety Regulation (GPSR) is a set of rules that requires consumer products to be safe and our obligations to ensure this.

If you have any concerns about our products, you can contact us on

ProductSafety@springernature.com

In case Publisher is established outside the EU, the EU authorized representative is:

Springer Nature Customer Service Center GmbH
Europaplatz 3
69115 Heidelberg, Germany

www.ingramcontent.com/pod-product-compliance
Lightning Source LLC
Chambersburg PA
CBHW070528020226
39081CB00003B/50
* 9 7 8 0 7 9 2 3 8 5 4 1 7 *